Author MELNICK, J.L.

MELNICK J.L. (ED.)
Progress in Medical Virology: Volum
e 28
LOC: QW 160/MEL 0001
~~0001971~~1

AF598557

Progress in Medical Virology, Vol. 28

Progress in Medical Virology

Vol. 28

Editor: *Joseph L. Melnick*, Houston, Tex.

Contributors
N.W. Axnick, Atlanta, Ga.; *J.R. Baringer*, San Francisco, Calif.; *R.F. Betts*, Rochester, N.Y.; *I. Hirsch*, Prague; *D.M. Knipe*, Boston, Mass.; *J.P. Koplan*, Atlanta, Ga.; *H.W. Lee*, Seoul; *H.C. Meissner*, Boston, Mass.; *J.L. Melnick*, Houston, Tex.; *W.G. Stroop*, San Francisco, Calif.; *P.C. Trexler*, Guildford; *G. van der Groen*, Antwerp; *V. Vonka*, Prague; *S.L. Wechsler*, Cincinnati, Ohio

9 figures, 1 color plate and 12 tables, 1982

S. Karger · Basel · München · Paris · London · New York · Sydney

Progress in Medical Virology

Vol. 22: X + 230 p., 24 fig., 16 tab., 1976, ISBN 3–8055–2315–7
Vol. 23: VIII + 212 p., 29 fig., 19 tab., 1977. ISBN 3–8055–2423–4
Vol. 24: VIII + 224 p., 5 fig., 37 tab., 1978. ISBN 3–8055–2810–0
Vol. 25: X + 180 p., 24 fig., 16 tab., 1979. ISBN 3–8055–2978–3
Vol. 26: VIII + 240 p., 58 fig., 1 cpl., 13 tab., 1980. ISBN 3–8055–0702–X
Vol. 27: Hepatitis B Virus and Primary Hepatocellular Carcinoma. Workshop on Hepatitis B Virus and Primary Hepatocellular Carcinoma, Dakar, Senegal, 1980. Ph. Maupas, Tours and J.L. Melnick, Houston, Tex. (eds.).
VIII + 212 p., 57 fig., 1 cpl., 64 tab., 1981. ISBN 3–8055–1784–X

National Library of Medicine, Cataloging in Publication
Progress in medical virology, v. 28, 1982, New York. v. illus.
ISSN 0079–645X.
1. Virology – yearbooks I. Title: Fortschritte der medizinischen Virusforschung
II. Title: Medical Virology III. Title: Progrès en virologie médicale
W 1 PR 6712
ISBN 3–8055–2983–X

Drug Dosage

The authors and the publisher have exerted every effort to ensure that drug selection and dosage set forth in this text are in accord with current recommendations and practice at the time of publication. However, in view of ongoing research, changes in government regulations, and the constant flow of information relating to drug therapy and drug reactions, the reader is urged to check the package insert for each drug for any change in indications and dosage and for added warnings and precautions. This is particularly important when the recommended agent is a new and/or infrequently employed drug.

All rights reserved

No part of this publication may be translated into other languages, reproduced or utilized in any form or by any means, electronic or mechanical, including photocopying, recording, microcopying, or by any information storage and retrieval system, without permission in writing from the publisher.

© Copyright 1982 by S. Karger AG, P.O. Box, CH-4009 Basel (Switzerland)
Printed in Switzerland by Thür AG, Offsetdruck, Pratteln
ISBN 3–8055–2983–X

Contents

Prog. med Virol., vol. 28, pp. 1–43 (Karger, Basel 1982)

Persistent, Slow and Latent Viral Infections

William G. Stroop, J. Richard Baringer

Departments of Medicine, Pathology and Neurology, University of California,
San Francisco, Calif., USA

Contents

I. Introduction

Several viruses from a wide variety of genera have been shown to persist in a susceptible host following either natural or experimental infection. This article will review some diseases of man and animals of which viral persistence is a feature, and diagram the variety of known mechanisms involved in the establishment and maintenance of persistence and the production of clinical disease. Moreover, where available evidence permits, new ideas concerning the pathogenesis of persistent viral infections will be offered. The viral infections of man and animal discussed illustrate the wide variation among host-virus relationships which can ultimately lead to morbidity and/or the persistent state. Persistent viral infections can be characterized by low or high levels of cell-free virus production, or an absence of complete virus maturation but with maintenance of either viral genomes within infected cells or limited expression of viral antigens. The spectrum of viral persistence may be present with or without production of obvious clinical disease of the host.

An example of a persistent viral infection of humans which is characterized by a high level of free virus expression and chronic disease production is progressive multifocal leukoencephalopathy (PML) [9, 77, 175]. Examples of persistent viral infections characterized by minimal expression of cell-free virus resulting in chronic progressive disease are progressive rubella panencephalitis (PRP), and subacute sclerosing panencephalitis (SSPE) [9]. Other persistent viral infections involve expression of little or no viral antigen in those organs or tissues harboring the virus, and cause only episodic periods of acute illness. Examples of viruses which cause these persistent infections in man include herpes simplex and varicella-zoster viruses [8].

Naturally occurring persistent viral infections of animals include visna and maedi in sheep [51, 103, 108, 175]; lactate dehydrogenase-elevating virus (LDV) [108, 120], lymphocytic choriomeningitis (LCM) virus [24, 108, 205], and Theiler's virus infections of mice [30–33, 93–100, 103, 177–179]; Aleutian mink disease [108, 140–143]; and equine infectious anemia [28, 112–116].

The diseases listed above are caused by conventional viruses, that is, viruses which have been shown to have generally typical structural, physicochemical and biological characteristics. Two progressive chronic disorders of man caused by unconventional agents, i.e., agents which have not been visualized, isolated or well characterized biologically or physicochemically, include Kuru and Creutzfeldt-Jakob disease (CJD) [9, 46, 77, 146, 147, 175]. Diseases of animals shown to involve persistent infection by unconventional agents include scrapie in sheep and goats, and transmissible mink encephalopathy (TME).

Within most of the viral animal-model systems the mechanisms of viral persistence are incompletely understood. Most acute viral infections are eliminated from the host by a concerted production of antibody to the infecting agent together with the activity of the cellular immune system. For virus to persist following an acute phase of replication in an immunocompetent host requires that it be able to evade normal immune surveillance and clearance mechanisms.

II. General Mechanisms of Persistence

Mechanisms for establishment and maintenance of viral persistence may be thought of as occurring under five general circumstances: (1) The agent may sequester itself in a 'protected site'. The most obvious are intracellular sites in which detection of the virus or viral antigens by the humoral or cellular immune system is prevented. Other 'protected sites' could be organ systems which are separated from the vascular compartment by an especially tight endothelial barrier, e.g., brain. Another 'protected site' in which viruses might persist is within cellular elements of the immune system. (2) Theoretically, a virus could persist if it was nonimmunogenic. (3) Viruses might persist if they elicited large quantities of non-neutralizing antibody which could block neutralizing antibody. Alternatively, non-neutralizing antibody-virus complexes could be phagocytosed by the reticuloendothelial system (RES) perhaps allowing the virus to reinfect the cells of the RES. (4) A virus might persist if it induces tolerance of the host; if the virus was relatively noncytocidal, the virus would be able to replicate freely without causing death of the host. (5) Viral persistence may result from growth of the agent in a less permissive cell type. A permissive cell is defined as one which allows complete expression of viral genetically coded functions and lytic production of free virus; a nonpermissive cell is defined as a cell which does not allow for complete viral expression.

The diseases discussed below illustrate these mechanisms of persistence. However, it would appear that no single mechanism is operative in any one disease; rather, multiple mechanisms of viral persistence seem to be involved in each disease, and each mechanism may contribute to the ultimate degree of morbidity of the host.

III. Persistent Infections Due to Conventional Agents

A. Progressive Multifocal Leukoencephalopathy (PML)

PML is an uncommon central nervous system (CNS) disease of humans and is due to an opportunistic papovavirus infection of patients with underlying disorders such as reticuloendothelial diseases, inflammatory diseases associated with a state of secondary immunodeficiency [9], Hodgkin's disease, lymphomas, leukemias, carcinoma and other malignancies, sarcoidosis, and of patients who have received immunosuppressive therapy [77, 175]. Patients with PML exhibit neurological signs indicating disease of the white matter, including paralysis, visual loss, sensory abnormalities, and ataxia. The disease is progressive and usually leads to death in less than 1 year. The demyelinative lesions of PML are characterized by enlarged oligodendrocytes with intranuclear eosinophilic inclusions containing viral particles, and large, bizarre dysplastic astrocytes. Inflammatory cells are usually not prominent in PML lesions [9], but have been observed in some cases.

Two papovaviruses, JC virus and SV40-PMLvirus [128, 129, 194, 195] have been recovered from PML brain. JC and SV40-PML viruses appear to be highly cell-associated; SV40-PML required fusion of explant cultures of PML brain with primary monkey kidney cells. JC virus was recovered by inoculating human-brain spongioblasts with PML brain material. Ultrastructural studies have shown that papovavirus particles are frequently found in oligodendroglial cells and rarely in astrocytes; immunofluorescence studies indicate more oligodendroglial cells bearing papovavirus T-antigen than astrocytes [175]. Oligodendroglial cells probably represent a permissive cell type which allows complete virus replication (fig. 1). Individuals who are immunosuppressed either through iatrogenic manipulation or a malignant disease process, and are infected with JC or SV40-PML viruses, are susceptible to a lytic infection of oligodendroglia (fig. 1). Serologic studies indicate that by 13 years of age, 65% of individuals have acquired antibody against JC virus, and at a later age, 72% of the population have humoral immunity to the agent [129], indicating a widespread subclinical infection. The PML agents, however, appear to display little virulence. It is still unresolved, however, if PML results from 'reactivation' of a persistent infection or whether it is due to a new infection.

The dysplastic astrocytes present in PML lesions suggest the possibility that the PML agents are capable of inducing transformation of these cell types. Transformation by the oncogenic DNA viruses and RNA tumor viruses usually occurs following infection of non-permissive cell types; infections of permissive cell types result in a lytic infection with release of viral progeny [14, 43, 189]. The transforming oncogenicity of JC virus has been demonstrated; JC virus produces a variety of brain tumors in hamsters after intracranial inoculation [129, 175].

Circumstantial evidence to support the postulate that astrocytes are nonpermissive to the PML agents has been obtained using in situ hybridiza-

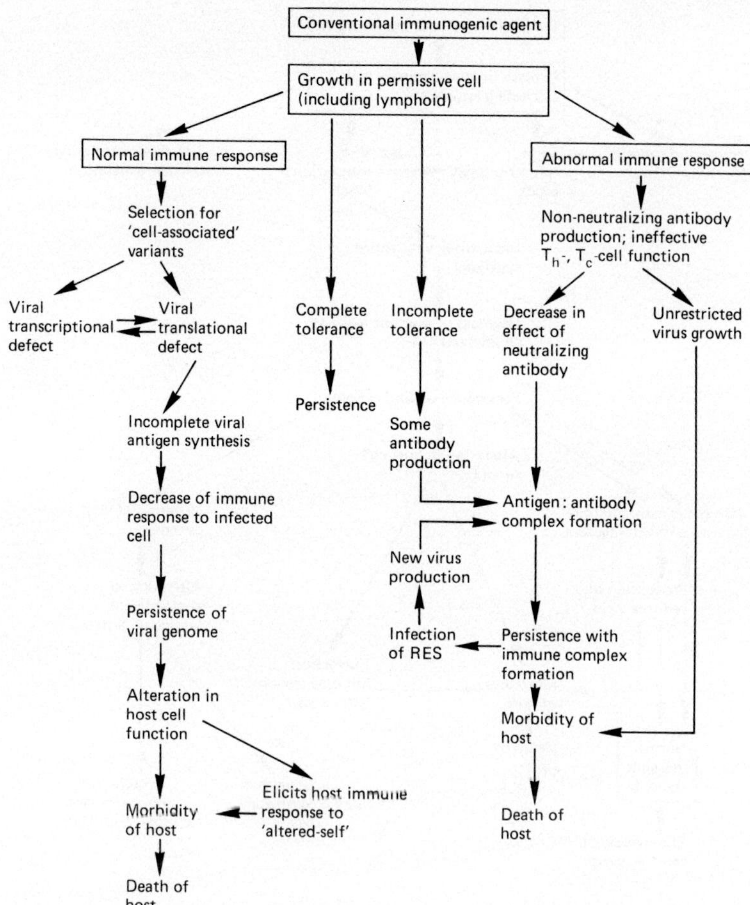

Fig. 1. Mechanisms of viral persistence and production of chronic disease due to infection of permissive cell types. The figure illustrates possible mechanisms of persistence of agents which can produce complete or incomplete tolerance of the host and which may induce normal or abnormal immune responses following infection. RES = Reticuloendothelial system. (See text for details.)

tion to localize JC virus DNA in PML brain [39]. These studies revealed heavy labeling of oligodendroglial nuclei but only a small amount of labeling of astrocytes within the lesions. These data do not prove or disprove the hypothesis of a nonpermissive infection of astrocytes, but at least indicate that astrocytes can be infected by the virus and are not exclusively reacting

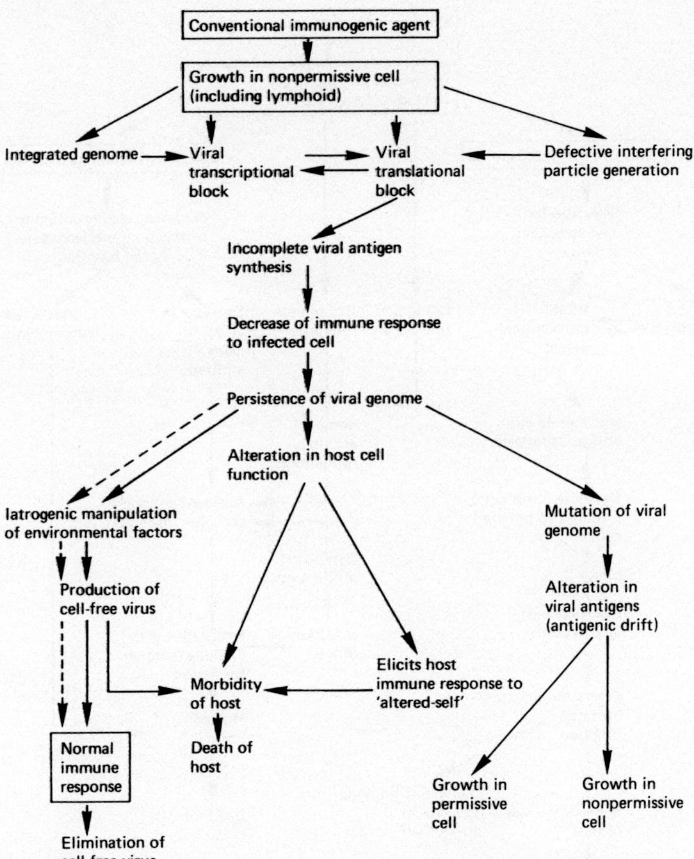

Fig. 2. Mechanisms of viral persistence and production of chronic disease due to infection of nonpermissive cell types. The figure illustrates possible mechanisms by which viruses may persist and produce morbidity of the host following infection of less-permissive cell types. The dashed line indicates the pathway of reactivation of a latent viral genome (e.g., herpes simplex) due to iatrogenic manipulation or environmental stimuli (see text for details).

to injury. If a nonpermissive infection of astrocytes does occur, such an infection may allow for partial persistence of viral genetic information within these cells (fig. 2). If the genome in nonpermissive cells was capable of partial gene expression, some synthesis of viral antigens may also occur; that immunofluorescence was observed in some astrocytes supports this hypothesis.

B. Subacute Sclerosing Panencephalitis (SSPE)

SSPE is a chronic progressive encephalitis of childhood or adolescence caused by measles virus [1, 3, 9, 16, 26, 34, 48, 73, 133, 155, 174, 176, 188]. The disease has its usual onset between 4 and 18 years of age, and appears to be associated most often with a natural measles infection of individuals under the age of 2 years. The disease is characterized clinically by subacute development of behavioral changes and dementia, progressing to development of myoclonic seizure activity, progressive obtundation into coma, and death [9].

Pathologically, SSPE brains are characterized by widespread lesions of both white and gray matter with relative sparing of the cerebellum. The lesions themselves are a mixture of demyelinative and necrotic foci with prominent astrocytosis. An inflammatory reaction is usually present consisting of lymphocytes and plasma cells. Eosinophilic nuclear inclusions are often present; occasionally, cytoplasmic inclusions may also be seen in neurons, oligodendroglia and astrocytes [9, 77]. The inclusions are composed of paramyxovirus-like nucleocapsids; immunofluorescent staining of brain sections has revealed measles antigens in both the nucleus and cytoplasm.

The host-virus relationship in SSPE may be thought of as one of virus replicating in a nonpermissive cell. This is suggested from the observation that conventional virus isolation techniques have usually failed to rescue the agent from brain material. Co-cultivation of trypsinized brain cells with cells permissive for measles virus is required to isolate the agent [131], but only succeeds in a minority of attempts. The brain isolates have been shown to differ in certain biologic and morphologic characteristics, but share antigens common to wild-type measles [77, 131]. SSPE isolates have been termed defective in that they most often fail to produce complete extracellular virus. Studies of the viruses recovered from diseased CNS tissue suggest that the SSPE agent is defective in its ability to produce the matrix (M) protein [54, 196] and also possibly the hemagglutinin (HA) protein [196], yet these SSPE isolates shared cross-reacting antigens with other measles viruses. Moreover, animal studies have indicated that multiple variants of measles virus were present in a stock of virus used for pathogenetic studies; one variant was capable of producing relapsing myelitis in hamsters [19]. Both variants, however, were serologically related [*Carrigan,* personal commun.].

Two theoretical mechanisms can be put forth to explain persistence of measles virus in SSPE. The first involves infection of an individual with a heterogeneous population of measles virus variants, the variants sharing antigenic similarities with each other and natural measles virus. The population of variants may undergo a selection process due to the normal immune response of the host (fig. 1). Evidence to support the role of the immune response as a selector for viral variants has been obtained with an animal model of SSPE. This model produces persistent CNS infection of hamsters and has indicated that the host antibody response to the virus seems to be involved in converting the virus from a cell-free to a cell-associated state [18, 75]. The cell-associated phenotype of these variants could be due to transcriptional and/or translational defects which cause incomplete viral antigen synthesis (fig. 1). Recent evidence by *Johnson* et al. [78] indicates that the M antigen selectively disappears during the course of infection in experimental SSPE in the hamster, at the time when serum antibody levels are increasing. Incomplete synthesis of M protein may decrease the antigen density of the virion glycoproteins (e.g., HA protein) on the surface of the infected cell, thus possibly allowing the cell to evade immune killing by cytotoxic T cells, by antibody-directed cell-cytotoxic mechanisms (ADCC), or by natural killer (NK) cells. The viral genome could thus persist in those cells not lysed by immune mechanisms (fig. 1).

The second mechanism involves infection of both permissive and less permissive cells with a homogeneous population of conventional measles virus during primary exposure (fig. 1, 2). The normal immune response would eliminate the free virus produced by permissive cells and those cells themselves. But the virus which infected less permissive cells could persist because the less permissive cell would not provide the specific or nonspecific factors required for normal transcription of viral nucleic acids or translation of its gene products. Perhaps the apparent absence of M protein in SSPE viruses reflects such a malfunction in normal complete gene expression.

Either mechanism would result in incomplete viral antigen synthesis, a subsequent decrease of the effectiveness of the immune response to the infected cell, thus fostering persistence of the viral genome (fig. 1, 2). Clearly, both mechanisms may be operative in SSPE. The persistence of the viral genome and limited expression of viral antigens may well serve to alter normal host cell function and result in cellular expression of antigens which are normally sequestered from the immune system. These 'neo-antigens', perhaps in association with minimal amounts of viral antigen, may elicit an autoimmune response to 'altered-self' (fig. 1, 2). *Panitch* et al. [130] have detected antibodies to myelin basic protein in the cerebrospinal fluid (CSF) of patients with SSPE, which may indicate that autoimmune mechanisms may be involved in the pathogenesis of SSPE. However, it is unclear whether the autoantibodies to myelin basic protein are specific or secondary to the disease process. Alterations in host cell function plus an elicitation of an autoimmune response may eventually lead to destruction of cells. Cumulative cellular damage will eventually lead to production of clinical disease, and death of the host (fig. 1, 2).

C. Progressive Rubella Panencephalitis (PRP)

Another chronic CNS viral infection with a clinical syndrome similar in some respects to SSPE is PRP [9, 77, 192, 197, 198]. The initial virus infection may occur either in utero, resulting in the congenital rubella syndrome, or as an acquired infection of childhood. In most cases, PRP has occurred in children with definite stigmata of congenital rubella. PRP is an extremely rare disease; as of 1980 only 7 cases were described. Of these 7 cases, 4 developed the disease as an apparent sequela to congenital rubella, 2 developed the disease with unknown primary exposure, and 1 with early childhood rubella [72, 182, 183]. PRP is thought to be due to a delayed reactivation of congenital rubella and acquired postnatal infections; it begins with progressive neurological impairment, including dementia, ataxia, spasticity and occasionally myoclonus, and leads to death after a protracted course of several years [182, 183].

PRP is an inflammatory disease characterized by perivascular cuffs of lymphocytes and microglial nodules within the brain parenchyma. Perivascular deposits of an amorphous periodic acid-Schiff-positive material are also present in PRP; this material is also frequently observed in brains of children suffering from congenital rubella syndrome. There are no inclusions present in brain tissue [9, 77]. In contrast with SSPE, the cerebellum is markedly affected and there appears to be greater white matter involvement in PRP than SSPE [183]. Rubella virus has been isolated from patients with PRP; *Cremer* et al. [29] were able to isolate rubella virus from brain biopsy material, and *Wolinsky* et al. [199] recovered the virus from peripheral blood leukocytes.

Cell-mediated and humoral immune responses have been examined in PRP and have failed to demonstrate a consistent defect which would explain the pathogenesis of the disease [29, 72, 157, 192]. In 2 cases, increasing serum levels of interferon were detected, and sera from these cases were shown to interfere with production of interferon by normal donor lymphocytes, in vitro [199]. Individuals with PRP have, however, large quantities of serum and CSF IgG [29, 182, 183, 198]; anti-rubella antibodies appear to be produced intrathecally [157].

Although it is probably premature to speculate on the persistence of rubella virus in PRP and the pathogenesis of the disease, available evidence suggests that some degree of host tolerance to rubella virus may be involved in viral persistence (fig. 1). Rubella virus infection in utero could produce a certain degree of tolerance to the virus; induction of tolerance to heterologous proteins during neonatal life has been demonstrated experimentally by *Habicht* et al. [53]. The hypothetical tolerance to rubella virus induced in utero in children with congenital rubella syndrome is supported by the observation that the virus is able to persist into childhood for considerable lengths of time [9, 105]. If the tolerance to rubella virus was complete resulting in total unresponsiveness of both T and B lymphocytes, no immunoglobulin to viral antigens would be produced, and no cell-mediated immune response to viral antigens would exist [191]. Because patients with congenital rubella and PRP do synthesize sometimes considerable amounts of virus-specific IgG, and in most instances show no major defects in cellular immunity, the hypothetical tolerance induced in utero must be incomplete (fig. 1). Continued replication and release of progeny virus coupled with at least partial immune unresponsiveness to the agent would provide for a situation in which free virus and membrane-associated viral antigen could complex with antibody and circulate as antigen:antibody complexes (fig. 1). Although partial tolerance may explain some features of rubella virus persistence in the congenital rubella syndrome, it is unclear if patients with PRP are to any degree tolerant to the agent. More detailed information about the immunology of this disease is needed before tolerance can be included or excluded as a mechanism of viral persistence in PRP.

Recent evidence suggests that immune complex formation occurs in the CSF and serum in PRP [27]. The immune complexes contained rubella-specific IgG as assayed by radioimmunoassay following removal of the immune complexes from serum or CSF by absorption to Raji cells. The density of the immune complexes varied from 1.14 to 1.26 g/cm^3. Preliminary studies of 1 case of PRP in our own laboratory indicate that immune complex deposition in vascular endothelia may contribute to the pathogenesis of the disease [184].

D. Murine Lymphocytic Choriomeningitis (LCM) Virus Infection

One of the best studied animal models in which tolerance of the host plays a crucial role in establishing a persistent viral infection accompanied by immune complex formation is LCM virus infection of mice. LCM is a member of the **Arenaviridae**. Because these viruses bud from infected cell membranes and are not particularly lytic, the pathogenesis of LCM infection is primarily immunopathologic in nature.

The clinical manifestation of LCM infection in mice depends on the strains of virus and mice used, route of infection, virus dose and, most interestingly, the immunocompetence of the host [24,37,205]. Totally immunoincompetent mice (fetal mice in utero, newborn mice less than 24 h old, nude mice, adult mice which have been thymectomized, lethally irradiated and bone-marrow-reconstituted, mice treated with cyclophosphamide at the time of or soon after infection, or mice treated with anti-lymphocyte serum), infected with LCM by the intracranial or intravenous routes, survive and become carriers of LCM virus. These mice lack virus-specific cytotoxic T lymphocytes, but produce non-neutralizing antibodies [108,205]. Production of antibodies to the virus may lead to immune complex disease later in life. The immune competence of newborn mice is more than that of fetal mice but less than that of adult mice. Intracranial or intravenous inoculation of newborn mice results in a mixture of clinical manifestations. Some mice may develop a 'runting' syndrome, some mice die from acute lymphocytic choriomeningitis, a few recover from this desease, and still others become chronic carriers of LCM virus [205]. When adult mice are infected with LCM virus by the intracranial or intravenous route, all mice die within 6–10 days after infection, of acute lymphocytic choriomeningitis [205].

The specificity of primed cytotoxic T lymphocytes is for viral antigens in conjunction with the major histocompatibility antigens (H-2) [206–210]. The generation of cytotoxic T lymphocytes has been demonstrated in a variety of virus-animal models in addition to LCM, including Sindbis [111], vaccinia [89], influenza [4], Sendai [38], coxsackie B3 [200,201] and CMV [65]. With the LCM, vaccinia, Sendai and coxsackie viral systems, primed cytotoxic T lymphocytes will lyse an infected cell target only if the cytotoxic T lymphocyte and the infected cell share homology in H-2 K or D antigens. The observation of H-2-restricted cytotoxic T-lymphocyte specificity made it possible to demonstrate that LCM virus infection of immunocompetent mice produces T-lymphocyte tolerance [24,205]. If tolerant, persistently infected mice are inoculated with syngeneic LCM-primed cytotoxic T lymphocytes, they will develop characteristic lesions [24,37,205].

Tolerance provides the mechanism for the establishment and maintenance of LCM virus persistence (Fig. 1). In this model, incomplete tolerance results in the production of some antibody. Infected cells are not lysed by cytotoxic T lymphocytes because of virus-induced tolerance. Mice become carriers of LCM virus because the virus itself is not particularly lytic, thus allowing host survival. Antigen:antibody complexes form in the circulation of tolerant infected mice and deposit in the glomeruli of the kidney; cumulative immune complex deposition may produce glomerulonephritis.

E. Lactate Dehydrogenase-Elevating Virus (LDV) Infection of Mice

Another model of persistence associated with late-onset glomerulonephritis is LDV infection of mice [125,138]. In contrast to LCM, most strains of LDV produce a life-long asymptomatic infection in mice of any age [109].

Although LDV has been known to be an infectious virus of mice since 1960 [150], it has been only recently that LDV has been classified as a togavirus [17]. LDV differs in its physicochemical properties from either the alpha- or flaviviruses of the **Togaviridae**. The agent was named for its ability to produce elevated levels of several plasma enzymes including lactate dehydrogenase, isocitric dehydrogenase, malic dehydrogenase, phosphohexose isomerase, glutamic oxaloacetic transaminase, and glutathione reductase [120,122]. The mechanism by which infection with most strains of LDV leads to elevated plasma enzyme levels appears to be due to an impaired clearance of endogenous enzymes from the peripheral circulation and is not due to leakage of enzymes from damaged cells or de novo enzyme synthesis [124]. The mechanism of impaired clearance of enzymes may be related to LDV infection of the RES.

LDV infection of mice produces widespread lymphoid hyperplasia and splenomegaly [108,162]. Interestingly, shortly after LDV infection, mice exhibit a transient decrease in the number of lymphocytes in the thymus, in the thymic-dependent areas of the spleen, and in the blood [162]. As with LCM infection, there is a persistent viremia associated with LDV infection. Infectious virus circulates as a complex with host IgG [109]. Immune complexes have been examined using an indirect immunofluorescence test; anti-LDV antibody has been detected in these complexes [108,143]. Not all of the immunoglobulins produced during LDV infection are neutralizing [108,123,143], as indicated by the presence of infectious virus:antibody complexes in serum of LDV-infected mice [143]; serum from infected mice is incapable of neutralization unless it is pretreated with ether or irradiated with ultraviolet light [123].

The immune complexes are filtered out of serum by the kidney and deposit on the epithelial side of the basement membrane of the glomeruli [125] in a manner identical to LCM virus immune complex deposition. Elution studies have shown that IgG from kidney contains anti-LDV specificity [125,143]. The immune complexes have been demonstrated to fix complement which elicits a mild proliferative glomerulonephritis. LDV antigens have also been detected in association with immune complexes [125,143].

The mechanism of persistence of LDV, although unknown, may be related to the interaction of virus with permissive cells, an apparent abnormal immune response, and circulation of infectious immune complexes (fig. 1). The virus is capable of replicating to relatively high titers in vivo without destroying infected cells. It is reasonable to postulate that an abnormal immune response to LDV, resulting in production of non-neutralizing antibody, may be related to the transient depression of T lymphocytes following infection with the agent; however, there is no direct evidence to support this postulate. The effect of partial T-lymphocyte depletion could be to decrease the number of helper T lymphocytes necessary for the ultimate synthesis of neutralizing antibody. *Notkins* [121] has suggested that disproportionate synthesis of non-neutralizing antibody could block the effectiveness of any neutralizing antibody produced.

The macrophage has been postulated to play a major role in viremic persistence by prolonging the animal's life through ingestion of immune complexes, thereby reducing the amount of immune complex deposition in the kidney [108]. Although LDV has been demonstrated by many investigators to be capable of replicating in a variety of phagocytic cells in vitro (primary mouse macrophages, mouse peritoneal macrophages, primary explants of reticuloendothelial cells) [17,42,137,204], no evidence exists which convincingly demonstrates the ability of LDV to replicate in these cells in vivo. By electron microscopy, LDV has been observed in peritoneal macrophages of infected mice [120], but it is unclear whether the virus was actually replicating in these cells or was simply phagocytosed. Because the virus has been shown capable of replication in macrophages in vitro, it is likely LDV can replicate in these cells in vivo.

Once macrophages have ingested the immune complexes, it has been postulated that the macrophage itself could become infected with the agent. Infection of the RES could have far-reaching effects on immune responsiveness to the agent as well as providing a reservoir for new virus production (fig. 1). It is well known that the macrophage is an important cell involved in stimulation of the immune response to T-dependent antigens [41, 45,81,135,152,153,160,186,187]; however, it is unknown whether LDV antigens are T-dependent or -independent. Infection of the macrophage may alter the cell's ability to process and present LDV-specific antigens to cooperating T lymphocytes in a manner compatible with normal immune responsiveness (fig. 1).

LDV has been shown to affect the functional capacity of the immune system. Under certain circumstances, LDV infection results in enhanced antibody production, depresses the induction of tolerance, depresses host-versus-graft and graft-versus-host reactions, and depresses phagocytosis [122]. Additionally, suggestive evidence has been obtained which indicates that LDV may induce immunosuppression of the host. It has been reported that certain tumors exhibit accelerated growth when the host is co-infected with LDV [185]. Other investigators have found that certain protozoan infections of animals are potentiated in their severity by LDV infection [122]. *Notkins* [122] has suggested that these effects could be explained on the basis of LDV-induced immunosuppression. It is possible that these diverse effects arise from LDV infection of macrophages. Clearly, LDV infection of the RES may produce many abnormalities in the host and possibly foster viral persistence. Careful examination of the immune response and the role of the macrophage during viral infection are needed to sort out the various factors involved in persistence of LDV.

LDV can also induce an age-dependent polioencephalomyelitis (ADPE) in C58 mice [104]. ADPE was initially discovered during studies of the immune response of C58 mice to the syngeneic leukemia I_b cell line. It was found that injection of inactivated I_b produced a paralytic CNS disease in mice immunosuppressed iatrogenically or by normal aging [40,110]. The I_b tumor line apparently harbored a particularly neurotropic strain of LDV. The LDV-like ADPE agent has been shown to replicate in mice of all ages. The mechanism of ADPE induction is currently unknown.

F. Theiler's Virus Infection of Mice

Theiler's murine encephalomyelitis viruses (TMEV) comprise a group of enteric viruses [126] which cause acute encephalomyelitis and limb paralysis after intracranial inoculation of mice [33,92,101,177–180]. TMEV has been recently characterized as a picornavirus [100,171]. There are two groups of TMEV: group II strains (the T0- and T0-like strains, including the DA, Yale, BeAn 8386 and WW strains) are poorly neuroinvasive following peripheral inoculation and are less virulent [33,126,179,180] than the group I strains (GD VII and FA strains). The strains of groups I and II are serologically cross-reactive with each other [93] but distinct from other picornaviruses and yellow fever virus [179,180].

Group II strains can produce a biphasic disease in weanling mice. The first phase is an acute polioencephalomyelitis during which virus replicates to high titers; mice which survive the acute infection uniformly develop a chronic, progressive demyelinating and/or necrotizing myelitis [33,92]. Group II-TMEV persist at relatively low levels in the CNS throughout the chronic phase despite the obvious inflammatory nature of the lesions and the high levels of serum neutralizing antibody [92]. From infected CNS tissue homogenates, persistent virus can be recovered which is capable of producing the biphasic disease when reinoculated into mice [92]. Group I strains do not cause demyelination and do not persist in vivo [94].

Brain-derived stocks of group II-TMEV are incapable of producing cytopathic effect (CPE) when inoculated directly onto a variety of tissue culture cells and require multiple passages in vitro until CPE develops [92, 93]. Tissue culture-adapted group II-TMEV are not capable of producing acute polioencephalomyelitis, but persist and produce demyelinating lesions following intracranial inoculation after a prolonged incubation period [94,99]. No such attenuation of group I strains occurs after tissue culture-adaptation [94]. It has been suggested that the differences in pathogenic potential between tissue culture-adapted group I and group II-TMEV depends on plaque size [94]; analysis of the polypeptides of group I and II-TMEV has revealed differences which have been interpreted to correlate with neurovirulence [100].

The chronic demyelinating phase of disease appears to be caused by mononuclear cell phagocytosis of myelin lamellae from axons [30–32]. The severity of demyelination has been shown to depend to some degree on the strain of mouse used [95,96]. There is evidence for recurrent remyelination during the demyelinative phase [32]. Neither viral particles within nor degenerative changes of CNS cells have been observed during the demyelinative phase [30]. Immunofluorescent studies have demonstrated the presence of viral antigens in neurons during the acute phase, but viral antigens have rarely, if ever, been observed during the chronic phase [92,96]. The paucity of antigens and absence of viral particles in the CNS during the chronic phase suggests that demyelination is immune-mediated. Evidence to support this concept has been obtained; demyelination has been prevented by immunosuppression [97,98], but polioencephalitis was increased by such treatments [98]. The potentiation of neuronal involvement correlates with an increase in viral antigens in these cells; immunosuppressed animals die of polioencephalomyelitis before they develop demyelinative disease [97,98].

Evidence has been obtained that implicates direct viral lysis of CNS cells as a mechanism of demyelination [132,203]. *Penney and Wolinsky* [132] demonstrated that the WW strain of TMEV (a group II strain) was capable of infecting the oligodendroglia of suckling mice during the acute phase of disease; ultrastructurally, crystalline arrays were observed in neurons and oligodendroglia [132], which probably correlate with the inclusions observed with this strain by light microscopy [202]. *Wroblewska* et al. [203] infected mouse CNS organotypic cultures in vitro with the WW strain and demonstrated that neurons, oligodendrocytes and astrocytes were infected.

The mechanism of the establishment and maintenance of TMEV persistence appears to involve infection of both permissive and less permissive cell types (fig. 1, 2), and the humoral and cellular immune responses may be involved in the pathogenesis of demyelination. In this model, the neuron appears highly permissive for viral replication, and glial cells, less permissive. Rapid viral replication would occur in permissive cells, resulting in high levels of free virus production. The ensuing inflammatory response produces the characteristic lesions of polioencephalomyelitis. Replication in less permissive cells yields lower amounts of virus, and perhaps allows for prolonged survival of the cell. Rapid clearance of the highly permissive cells through neuronophagia by inflammatory cells, and neutralization of free virus by antibody, would tend to decrease the number of infected permissive cells. The less permissively infected cells would represent the sites of viral persistence.

The cellular immune response would, as a result, be 'shifted' from permissive 'gray matter' cells to less permissive 'white matter' cells, accounting for the change in lesion distribution throughout the course of infection. Continued replication of the virus in glial cells with antibody present in the extracellular milieu prevents widespread virus dissemination; only through cell-to-cell spread could progeny virus be capable of establishing new sites of infection. Eventual lysis of infected glial cells could give rise to primary demyelination; the cellular immune response to these infected cells could give rise to secondary immune-mediated demyelination.

This model, termed the 'relative-permissive' model of Theiler's virus persistence, accounts for the low levels of persistent cell-free virus present during the chronic phase of disease and the shift of lesion distribution; it also offers an explanation for the paucity of viral antigen and the failure to detect viral crystalline arrays during the chronic phase. Cells that are highly permissive contain large crystalline arrays whereas less permissive cells probably would not; moreover, there are probably fewer less-permissive infected cells during chronic disease than infected permissive cells during the acute phase. Furthermore, this model predicts that the effect of immunosuppression would be to allow spread of virus to permissive cells by limiting the amount of neutralizing antibody present in the extracellular milieu. Neuronolytic infection would ensue, potentiating gray matter disease; the immunosuppression experiments mentioned above are consistent with this concept.

The site of viral replication of the tissue culture-adapted group II strains would appear to be restricted to white matter. The apparently white-matter-exclusive tropism of tissue culture-adapted strains as opposed to the dual, gray and white matter tropisms of brain-derived strains could be explained by two mechanisms. Either two variants are present in brain-derived stocks, and the gray matter-tropic (e.g., encephalitogenic) variant is selected against by tissue culture adaptation leaving only the white-matter-tropic variant (e.g., demyelinogenic), or a mutation occurs during the adaptation process producing virus capable of replication only in glial cells. If a selection process occurs in vivo during the disease course, it must not be as complete as the selection in vitro, as virus recovered from persistently infected animals retains its encephalitogenic potential [92]. If a mutation event is responsible for the attenuated pathologic potential of these strains, it would most likely occur in the same region of the viral genome for each group II strain adapted to tissue culture; presumably the mutations would occur near the 5′- end of the genome, within the capsid polypeptide coding region, since mutations in the 3′- end, within the enzymatic and replicase polypeptide coding region, would most likely be lethal. Generation of poliovirus deletion mutants during high multiplicity passage in vitro has been studied [102]. In this system, the viable mutants contained deletions in the 5′- end of the genome.

Although the available evidence suggests that the pathogenesis and persistence of TMEV infection may involve CNS cells with different levels of permissiveness for viral replication, more experiments are required to elucidate the precise cell-virus interaction during the course of disease and to define the relationship between viral persistence and demyelination. These experiments should be aided by the use of the purely demyelinogenic tissue culture-adapted group II TMEV strains.

G. Aleutian Mink Disease (AMD)

Aleutian mink disease (AMD) is a chronic disease of mink in which Aleutian mink disease virus (AMDV) persists. AMD was first recognized to be caused by a virus in 1962 [82]. AMDV has been recently shown to have the biochemical and biophysical characteristics of a parvovirus [15]; the DNA of AMDV is of negative polarity [15], suggesting that it is a member of the autonomous group of parvoviruses, like minute-virus of mice. Vertical and horizontal transmission of the agent has been demonstrated and once mink are infected, the virus persists for life [71].

AMD is a chronic, slowly progressive condition characterized by glomerulonephritis, arteritis, occasionally focal hepatitis, and – most predominantly – a systemic proliferation of plasma cells resulting in a marked hypergammaglobulinemia [141]. The hypergammaglobulinemia may progress to a monoclonal gammopathy which is secondary to the plasma cell proliferation and is directly due to an overproduction of IgG rather than a defect in catabolism [141]. Death may occur from 2 to 24 months after infection. The genetic constitution of the mink influences the onset and severity of disease and eventual mortality; mink which are homozygous recessive for the Aleutian coat color genes develop particularly severe disease [58,109,141]. Although anti-AMDV antibody is present in high levels in serum, the antibody is incapable of neutralization. Glomerular and arterial wall lesions have been demonstrated to contain deposits of IgG and the C3 component of complement; the IgG from kidney eluates has AMDV specificity, and AMDV antigen has been demonstrated in acid eluates from arterial walls [60,141,142]. AMD bears pathogenic and immunologic similarities to LCM infection and LDV infections of mice, i.e., production of neutralizing and non-neutralizing antibodies, immune complex formation and eventual morbidity of the host due to glomerulonephritis (fig. 1).

The hypergammaglobulinemia characteristic of AMD represents an extreme humoral response to AMDV antigens [141], but is insufficient to neutralize the virus. *Porter* and his colleagues [141] have suggested that the macrophage plays a key role in virus persistence in AMD. In this model, the macrophage is thought to ingest circulating virus:antibody complexes and to become infected with the virus; replication within the macrophage is thought to result in renewed virus production. The progeny virus can then complex with virus-specific antibody and deposit in the glomeruli and upon the intima of blood vessels, triggering glomerulonephritis and vasculitis.

The hypergammaglobulinemia further suggests that AMDV infection results in a severe breakdown of normal immune regulation. It is conceivable that, if AMDV is capable of replicating in macrophages, normal antigen processing and presentation may be altered. Abnormal regulation of the immune response to AMDV may produce abnormal clonal expansion of those plasma cells which secrete specific anti-AMDV antibody. Inappropriate immune regulation is further suggested by the observation of several characteristics of the disease similar to autoimmune disorders of man, including the presence of Coombs-positive hemolytic anemia, LE cells, anti-DNA antibody and antinuclear factor, in addition to immune complex glomerulonephritis [91].

Although an attractive model for the mechanism of AMDV persistence and disease production, no conclusive evidence has been obtained that AMDV is capable of replication in macrophages in vivo or in vitro. Immunofluorescence studies have demonstrated that AMDV antigens are present in the macrophage in vivo [141]; however, this evidence alone is insufficient to prove that the virus is replicating in this cell type; positive fluorescence could well be due to phagocytosed antigen:antibody complexes. More recent studies by *Hensen* et al. [62] demonstrated positive fluorescence mostly in the cytoplasm but also in the nuclei of cells in the cortical areas of lymph node and the perisinusoidal cells of the liver. These investigators were not able to determine whether the cytoplasmic fluorescence in these cells represented phagocytosed antigen or antigen being produced within these cells. Nuclear fluorescence has been observed in cat cells infected in vitro [22]. The nuclear fluorescence is consistent with a nuclear dependent phase of replication of a parvovirus. Further experiments are needed to resolve the questions of whether AMDV can replicate in macrophages in vivo and the mechanism of the virus-nuclear dependent phase, in order to define the mechanistic role of the macrophage in the persistence of AMDV.

IV. Persistent Latent Viral Infections

A. Equine Infectious Anemia (EIA)

EIA and visna (see below) are two additional diseases in which viral persistence is a feature. EIA virus (EIAV) is a retrovirus [28] which produces a persistent viremia in horses [60]. Infected horses develop an acute illness 1–2 weeks after primary exposure. Of those animals infected, some die early in the course of disease, and others recover from acute illness only to subsequently suffer recurring disease months or years later which may lead to death. A few infected horses become asymptomatic carriers of the virus.

Active EIA is characterized by severe hemolytic anemia, widespread lymphoproliferative lesions, hypergammaglobulinemia, hepatitis and glomerulonephritis [60]. Viremia occurs in the presence of neutralizing and complement-fixing antibodies; some of this antibody is present in the form of immune complexes [60,115,116]. The glomerulonephritis is due to deposition of immune complexes as evidenced by deposits of IgG with anti-EIAV activity and C3 present in the glomeruli of infected horses [6]. The glomerulonephritis probably follows an immunopathological mechanism analogous to LDV infection of mice and AMD. The hemolytic anemia for which EIAV was named is due to a complement-dependent lysis of erythrocytes [112,113]. Binding of viral antibody to the surface of erythrocytes leads to complement fixation and lysis [61,91,112]. The targets on red blood cells to which anti-EIA antibodies attach are presumably the hemagglutinin subunits of the virus [28]. During active hemolytic disease, horses have reduced levels of complement; the depressed levels of circulating complement are consistent with the presence of the infectious immune complexes, glomerulonephritis, and erythrocytolysis. Hepatitis in EIAV infection appears to be caused by cell-mediated lysis of infected hepatocytes [60,91]; depression of cell-mediated immunity by cyclophosphamide treatment ablates immunopathologic hepatocytolysis [61,91].

The role of the macrophage in persistence of EIAV is probably better defined than in either LDV infection of mice or AMD. The early experiments of *McGuire* et al. [114] demonstrated that specific EIAV fluorescence could be observed in tissues following experimental inoculation of horses with EIAV. Viral antigens were observed in a variety of tissues including the thymus, liver, kidney, bone marrow, lung, stomach, intestines, adrenal glands, pancreas, cerebrum, cerebellum, the red but rarely the white pulp of spleen, and the medullary areas of lymph nodes. Where they could determine in which cell fluorescence was observed, the macrophage was the most common. This was especially prominent in the liver, in which positive fluorescence was easily observed in Kupffer cells but not in adjacent hepatocytes. The time course of appearance of specific fluorescence gave these investigators the impression that the macrophage was the first cell to exhibit positive fluorescence, and therefore the first cell type to support viral replication. The appearance of positive fluorescence occurred prior to the onset of viremia, indicating that presence of positive fluorescence in these cells was not due to phagocytosis of immune complexes.

Perhaps the most unique feature of EIA is that many persistently infected horses recover from the initial acute illness, but subsequently suffer from recurring episodes of clinical illness. *Kono* et al. [88] have shown that each successive clinical episode is accompanied by virus differing slightly in antigenicity from the original infecting agent and from viruses isolated from previous episodes [28, 88]. This process has been termed 'antigenic drift', and may provide a further mechanism of viral persistence (fig. 2).

Direct evidence has been obtained indicating that EIAV preferentially replicates in macrophages [28]. Field isolates of wild-type EIAV will replicate well in horse macrophages in vitro but only very poorly in other equine cell types, and not at all in cells from non-equine species. In a natural infection, the virus would almost certainly replicate in a permissive cell (macrophage) and not in less permissive cells. Thus, pathogenetic considerations, plus available experimental evidence, strongly suggest that the macrophage is probably the primary site of replication in vivo, but infection and replication in other cell types may also occur.

EIA may involve infection of both permissive and nonpermissive cells. In the permissive cell, normal maturation occurs, and progeny viruses bud from the cell surface. In the less permissive cell, the proviral genome may be integrated. However, the lowered permissivity of the cell may result from either a transcriptional or translational block in the path of virus maturation; intracellular restrictions on retroviral replication have been shown to exist, even though the complete viral genome may be integrated [14, 189, 193]. Such intracellular restrictions may cause incomplete viral antigen synthesis, which, in turn, decreases the effectiveness of the immune response to the infected cell. Presence of the integrated genome in a less permissive cell allows persistence of viral genetic information within the host (fig. 2). Following lytic growth of EIAV in the permissive cell (macrophage), the host mounts a normal immune response; immune complexes are formed as a result of the vigorous humoral response. Infection of the RES may result from phagocytosis of infectious immune complexes. Renewed viral replication in these permissive cells would result in the release of progeny virions bearing the same antigenic specificities as the original infecting virus. Virus could persist for a limited time, but only until the normal immune response has cleared the infection. Some infected horses die of immune complex disease during this acute phase; however, some recover and subsequently develop episodic recrudescences characterized by production of virus with different antigenic specificities. The integrated genome present in the less permissive cell may provide a source for new virus production.

Antigenic drift of EIAV may occur if the provirus present in the less permissive cell undergoes mutation. The mutations must occur in the genes necessary for viral replication and also in the genes coding for the major envelope glycoproteins to produce virus antigenically distinct from the initial infecting virus. Antigenically distinct progeny could reinfect new permissive and less permissive cells. Infection of permissive cells would produce renewed active disease; infection of less permissive cells could provide new sites of viral persistence.

Evidence has been obtained indicating that EIAV can establish a persistent infection of less permissive cells. *Crawford* et al. [28] established a persistent infection of equine fibroblasts and characterized the production of EIAV and EIAV antigens during the course of infection. However, these studies did not assess whether a lysogenous infection was directly involved in the pathogenesis of disease. Further analysis of the virus-host interaction in EIAV infection in vivo and in vitro will be necessary to completely sort out the mechanisms of persistence, and the generation of the variants.

It is unclear what role antigenic variants of EIAV play in the episodic nature of EIA. The presence of a new antigenic variant in the blood in concert with each new clinical cycle permits a temporary escape of the new variant from the immune system, 'primed' by the previous variant. The appearance of antibody to the new variant in a few days apparently re-establishes immunologic control, and the viremia subsides. Because little antigenic relatedness exists between the viral isolates [28,88], cross-protection by antibody is insufficient to prevent reinfection during each new clinical cycle. The mechanisms responsible for the eventual cessation of the clinical cycles seen in most horses are not known.

B. Visna Virus Infections

Another persistent viral disease of animals in which antigenic drift has been demonstrated is visna-maedi of sheep [117,158]. In 1954, *Sigurdsson* [161] introduced the term 'slow virus infections' to describe the unusual characteristics of several diseases that appeared in Iceland during 1930–1950; these diseases included visna-maedi. Visna ('wasting') and maedi ('shortness of breath') refer to the neurological and pulmonary manifestations, respectively, of the infection. Visna-maedi has appeared elsewhere in the world and is known as Zwoegerziekte in the Netherlands and as progressive pneumonia in the United States.

The viruses that cause visna belong to the Lentivirinae subfamily of the **Retroviridae.** The lentiviruses, unlike the oncornaviruses, are nontransforming and are not associated with neoplasms but instead cause slowly progressive inflammatory diseases. The lentiviruses are serologically distinct for both type- and group-specific antigens from the other retroviruses and EIAV [170] and, further, do not share any nucleotide sequence homology with them [55]. Lentiviruses have a mechanism of replication similar to but not identical with that of the other members of the **Retroviridae;** however, replication does not require cellular DNA synthesis or mitosis.

In nature, visna virus appears to be transmitted horizontally from animal to animal; lambs receive virus from ewes via the milk, older animals transmit the virus by respiratory aerosols or infected saliva. Since endogenous lentivirus genes have not been detected in uninfected cells [50], and transplacental transmission has not been detectable experimentally [35], it is unlikely that vertical transmission occurs.

Experimental infections of sheep have done much to elucidate the pathogenesis of visna. Following intracranial inoculation, the virus replicates in the choroid plexus and ependymal cells of the CNS. A viremia occurs which disseminates virus to distant organ systems including the RES and the lung. Inflammatory cells infiltrate the infected areas a few days after inoculation, and the cellular immune response peaks at about 2 weeks, decreasing in about 1 month [49]. The inflammatory response to infection of the lungs produces dyspnea and progressive pneumonia. Complement-fixing antibodies appear a few weeks after inoculation, rise to maximum titers in a few months, and remain elevated throughout the course of the disease. Neutralizing antibodies appear later than complement-fixing antibodies (between 1 and 4 months postinoculation) and also remain at high levels [134]. There are also nonimmunoglobulin components in serum and cerebrospinal fluid capable of neutralization [181].

The incubation period of experimentally infected sheep is irregular, ranging from 2 months to over 10 years; the clinical phase follows a protracted course, generally extending over several years. The onset of the disease is insidious. Slight aberration of gait, especially of the hind quarters, unnatural tilting of the head and, in rare cases, blindness may be observed. The disease progresses to hemiparesis or total paralysis; unattended animals will die of malnutrition.

Visna virus replicates poorly in experimentally inoculated fetal lambs. *Narayan* et al. [119] found that recovery of virus from choroid plexus, brain, lung, spleen, kidney and the buffy coat was much more efficient from tissue explants than from tissue homogenates. Recovery of virus was possible for weeks after infection, although the frequency of successful virus recovery from explant cultures decreased after the first 2 months post-infection. When similar studies were performed using adult sheep, virus could be recovered from tissue explants for 1 year after intracerebral inoculation; however, no extracellular virus was found between 10 days and 1 year post-inoculation [119]. In contrast, visna virus causes lytic, productive infection in sheep choroid plexus in vitro [51, 52]. Thus it would appear that visna virus replication in vivo is restricted, compared to virus replication in vitro. Furthermore, despite the vigorous inflammatory and humoral immune responses, visna virus is capable of persisting in vivo throughout the course of infection.

The nature of visna virus persistence in vivo would appear to involve a virus-host cell interaction in which the provirus is maintained within a certain number of cells that can evade the immune system. *Haase* et al. [51,52] have found that many cells of the choroid plexus of an animal infected with 10^7 PFU of virus contained viral DNA, as detected by in situ hybridization, at about the same proportion as scored by infectious foci experiments [51]. Only a small fraction of cells containing viral DNA were found to contain viral-core-specific antigen (P30) and major virion glycoprotein antigen (gP 135) [52]. *Haase* et al. [52] have developed a quantitative method for determining the amount of visna viral RNA and DNA per cell. During the course of productive lytic infection in vitro, viral RNA increased from a few copies per cell to several thousand copies per cell by the end of the growth cycle [52]. Only a few cells in sheep choroid plexus from infected sheep contained viral RNA, but many cells contained viral DNA. These data suggest that viral DNA is maintained in cells in vivo but expression of viral genetic information is blocked at the transcriptional level (fig. 2). Those cells capable of expressing viral RNA may be the source of the small amount of free virus found in infected tissues [52]. The few cells in infected animals which express viral RNA could be thought of as 'productively infected' and would be the targets for immune cytolysis. The vast majority of cells, however, maintaining the viral genome in a lysogenous or latent state, would be immunologically 'silent' and would therefore be unrecognized by the immune system (fig. 2).

Narayan et al. [119] demonstrated that peripheral blood leukocytes (PBLs) of infected sheep contained visna virus DNA. When PBLs were inoculated with visna virus no lytic replication was detected; however, virus could be recovered from these cells by co-cultivation techniques [119]. It is possible, therefore, that latently infected PBLs may harbor proviral visna DNA and serve to disseminate virus throughout the animal. For virus to infect other tissues from a latently infected PBL requires that the restriction on viral gene expression be lifted, so that the cell can release progeny virus into its immediate environment. No direct evidence has yet been reported indicating that the latently infected PBL plays this role in the pathogenesis of visna. Alternatively, the PBL may be chronically infected, releasing progeny below the level of detection. Such chronically infected PBLs could serve as harbingers for viral dissemination.

Antigenic drift has also been shown to occur in visna [117–119]. Following intracerebral inoculation of a sheep with one strain of visna virus, three isolates of virus were obtained from PBLs during the 3-year course of infection. The isolates were compared with the parental virus by mapping of T_1-resistant oligonucleotides [23]. The three isolates were antigenically distinct from the parental virus and from each other [118] and were found to contain mutations in the gene that codes for the antigen to which the neutralizing antibody was directed [23, 118, 158]. Interestingly, although antibodies to the parental virus and the three antigenic variants of visna virus appeared to develop in sequential order [118,119], the parental and variant strains could be isolated contemporaneously from PBLs.

A theoretical mechanism for the pathogenesis and persistence of visna virus is outlined in figure 2. Infection of a less-permissive cell type may result in the establishment of lysogeny. Complete viral expression is limited, resulting in incomplete mRNA synthesis, consequential incomplete synthesis of viral antigens, and a decrease or an absence of an immune response to lysogenously infected cells; only those permissive cells expressing viral antigens are attacked by the immune system. Mutation of the 3′- end of the genome may result in alteration of the major glycoprotein of the virus. Virus released from such cells will not be neutralized by antibody directed to the parental virus; the antigenic variants may then infect other less permissive and permissive cell types.

The role of antigenic variants in the persistence of visna virus, dissemination of virus to distant organs, and the evolution of chronic progressive disease is unclear for two reasons. First, all visna virus variants were derived from PBLs explanted in culture, indicating that the variant strains were not found extracellularly. Second, unlike the antigenic variants of EIAV, visna variants do not replace one another throughout the course of infection. As it is clear that the parental virus is capable of persisting in the host despite the generation of viral variants in the face of a vigorous immune response, the role of the variants in the maintenance of persistence is unclear. It is perhaps conceivable that the variants may play a role in spread of visna virus throughout a population of sheep in a manner reminiscent of the epidemiology of influenza epidemics, although there is no evidence to support this postulate. Clearly, further investigations are necessary to determine what role, if any, antigenic variants of visna virus play in the pathogenesis of visna, the maintenance and establishment of viral persistence in a single animal, and the relationships between the immune response and the lysogenously and lytically infected cells.

C. Herpesvirus Infections

In addition to the lentiviruses, the herpesviruses are capable of establishing latent persistent infections in vivo. Of all members of the genus *Herpesvirus*, Epstein-Barr virus (EBV) and herpes simplex viruses types 1 and 2 (HSV-1, HSV-2) have been the most studied with respect to their ability to induce latency, to persist, and to establish the relationship between latency and the appearance of disease. HSV is responsible for recurrent infections in man, most often taking the form of periodic cutaneous, oral or genital mucosal infections, but the virus has also been shown to be responsible for serious generalized infections in newborns or to produce meningitis, myelitis or encephalitis in adults [8]. EBV is associated with infectious mononucleosis, African Burkitt's lymphoma, nasopharyngeal carcinoma [56,57], and some cases of Guillain-Barré syndrome [75].

Although HSV is capable of lytic replication in a wide variety of cell types in vivo and in vitro, the sensory and autonomic ganglia of animals and man are the sites where HSV has been demonstrated to persist in a latent form [11,13,145,165–167]. EBV preferentially infects B lymphocytes in vivo [106] and in vitro [107].

Much of the information on the pathogenesis of HSV infections has been obtained from animal models. In experimental animals following inoculation by various routes, the virus spreads centripetally within axons [10, 25, 79] to infect the appropriate dorsal or autonomic ganglion [8, 145]. The acute stage of ganglionic infection is defined by the presence of infectious virus in cell-free ganglion homogenates and relatively easy visualization of viral particles by electron microscopy within the ganglion [8]. After the productive infection subsides the chronic or latent stage is established; during this stage the virus can be demonstrated only by explantation of ganglia or by co-cultivation with suitable indicator cells [7, 8, 11, 13, 167, 168]. *Walz* et al. [190] have shown by infectious center assays that the fraction of ganglion cells harboring the virus decreases from 1% during the acute stage to 0.1% during the chronic stage.

The latent ganglionic infection can be induced to produce progeny virus by explantation of latently infected ganglia in vitro [8]. Reactivation of a latent ganglionic infection resulting in the release of progeny virions at a peripheral site has been achieved. *Stevens* et al. [169] were able to reactivate a latent sacrosciatic ganglionic infection by intratracheal instillation of *Diplococcus pneumoniae*; virus appeared to travel centrifugally and centripetally along nerves. *Hill* et al. [63] have developed a recurrent HSV ear infection of mice; trauma to the ear can produce reactivation of the virus at the site of inoculation and a recurrence of the lesion [64]. *Openshaw* et al. [127] have demonstrated reactivation in vivo in the presence of antibody following irradiation or cyclophosphamide treatment of latently infected immunocompetent mice. These studies indicate that the host immune response plays a critical role in controlling the acute phase of ganglionic infection.

The stereotypic recurrent mucocutaneous lesions produced by HSV are in all likelihood due to reactivation of the latent ganglionic infection. A variety of environmental factors have been demonstrated to provoke recrudescences (fig. 2); these include exposure to sunlight, allergic reaction, trauma, menstruation and fever. Iatrogenic manipulation of the centrally projecting sensory fibers of the trigeminal ganglia has also been demonstrated to provoke HSV lesions [20, 21] (fig. 2). It is unclear whether reactivation of a latent HSV infection or a new infection is responsible for herpes encephalitis in adults.

Although little is known of the virus-host relationship during the acute phase leading to the establishment of latency, even less is known about the state of the latent viral genome in infected ganglion cells. The central unsettled question in this system is the state and activity of the latent genome. Two theories have been put forth to describe the activity of the genome, 'static' and 'dynamic' [148, 151, 165]. Static latency is a condition in which the genome is thought to be maintained in a non-expressive state. Dynamic latency is a state in which the virus is envisioned to multiply at a slow rate, producing infectious virus at levels that cannot be detected by standard techniques. The latter is reminiscent of visna virus infection of sheep choroid plexus, outlined above.

Recently, several groups of investigators have examined acute and latent HSV infections in vivo to determine the level of viral specific DNA and mRNA present in cells. *Puga* et al. [148] followed the reassociation kinetics of acutely and latently infected ganglionic DNA and RNA mixed with ^{125}I-labeled viral DNA. During the acute phase HSV was recovered from ganglionic homogenates. Viral DNA was present in acutely infected cells at a level of 1.2–2 genome equivalents per cell, and at 0.11 equivalents per cell in latently infected cells; viral specific mRNA was found at a level of 0.1–0.2

equivalents per cell during acute infection but at less than 1 equivalent per 2,000 cells in latently infected ganglia. *Galloway* et al. [47] used a nick translated ^{3}H-DNA probe specific for HSV-2 RNA to examine the amount of RNA present in paravertebral ganglia, removed from humans at autopsy, by in situ hybridization. These investigators found that only 0.4–8% of the neurons present in the ganglia were expressing HSV RNA; the RNA detected was in association with the Nissl bodies, indicating that the neurons were elaborating HSV-specific RNA. ^{3}H–DNA probes specific for viruses other than HSV did not hybridize to any ganglion cells.

These results indicate either that many latently infected ganglion cells contain HSV viral DNA and only some cells express viral genes, or that only some neurons are latently infected. In either case, it would appear that the latent state involves a host-virus interaction which is restrictive for complete viral mRNA transcription (fig. 2). By analogy to visna virus infection of sheep choroid plexus, one prediction of the transcriptional block would be that few or no viral antigens should be expressed in latently infected ganglia.

Levine et al. [90] established a persistent HSV-1 infection in vitro using B 103 neuroma cell lines. Following a brief lytic infection, surviving cells ceased to produce progeny and appeared identical to uninfected cells. The persistently infected neuroma cells elaborated certain HSV-1 antigens, but not all of them. The antigens responsible for complement-mediated immune cytolysis were not expressed in the persistently infected cell line (vis-à-vis productively infected cells). Absence of these antigens provided resistance to complement-mediated immune cytolysis in vitro [90]. Limited expression of viral antigens during the latent phase in vivo has been demonstrated by one study. *Rajcani* et al. [149] examined established latently infected ganglia for the presence of HSV antigens. Using antisera directed against early and immediate-early HSV antigens (antigens associated with early events in replication), no positive immunofluorescence was observed in serial sections of latently infected ganglia from rabbits. After ganglia were explanted in culture, however, positive fluorescence of these antigens was detected; appearance of immediate-early and early antigens apparently correlated with 'explant-reactivation' [149]. Although these data are unconfirmed, they indicate that HSV infections involve no expression of viral antigens during latency.

Available evidence suggests that maintenance of persistent latent HSV infections in vivo is associated with a transcriptional (and subsequent translational) block (fig. 2); however, little information is available which sheds light on the establishment of latency.

Tenser and Dunstan [172] and *Tenser* et al. [173] have suggested that HSV thymidine kinase (TK) expression may be required for establishment of latent ganglionic infection. Rabbits were infected with TK^+ and TK^- strains of HSV by the cornea scarification technique [12]. HSV TK^+ strains were capable of replication in the cornea and the trigeminal ganglia, whereas HSV TK^- strains were unable to replicate in the ganglia, but could replicate in the cornea. These data indicate that the expression of TK is required for acute replication in the ganglia. *Fong and Scriba* [44] used ^{125}I-deoxycytidine as a substrate to measure TK activity in latently and acutely infected ganglia of guinea pigs; they could not detect TK during latent infection. It is possible that TK is required for establishment of latency, but once established, the TK gene need no longer function.

The host's immune response during acute ganglionic infection has been implicated by *Sokawa* et al. [163] to play a role in the conversion from the acute to the latent phase of infection. Ganglia were examined during the acute and latent stages of infection for the presence of 2′5′-oligoadenylate synthetase and interferon activities. Interferon has been demonstrated to induce cells to produce at least two enzymes: one is 2′5′-oligoadenylate synthetase, which catalyzes the formation of 2′5′-oligoadenylate from ATP in the presence of double-stranded RNA; the other enzyme is a double-stranded RNA-dependent protein kinase, which inactivates initiation factor eIf-2 [5]. Their data indicated that the acute phase of HSV replication in infected ganglia was suppressed by interferon production; the suppression occurred before antibody synthesis had begun [163]. Although unconfirmed, this report indicated that the host's immune response may play a role in establishment of latency. More information is required to define the role of the immune system during HSV infections to establish its importance in modulating, altering or regulating the ultimate degree of HSV expression.

V. Slow Virus Infections Caused by Unconventional Agents

Kuru, Creutzfeldt-Jakob disease (CJD), scrapie, and transmissible mink encephalopathy (TME) are diseases caused by unconventional infectious agents. The clinical, immunological, and molecular biological aspects of the diseases caused by these agents have been recently reviewed [146,147]. The agents responsible for these diseases are unusually resistant to a variety of reagents which inactivate conventional viruses, including formaldehyde, β-propiolactone, trypsin, pepsin, ribonucleases A and III, and deoxyribonuclease I [46]. They have not been visualized by electron microscopy in infected tissues, and purification of the agents has not yet been completely successful. The agents have an unusually prolonged incubation period and have been termed 'unconventional slow viruses' [46, 161]. In their respective hosts, they produce chronic progressive degenerative CNS diseases which are uniformly fatal [46, 146, 147]. These agents produce a progressive vacuolation in the dendritic and axonal processes and cell bodies of neurons; the end stage of the pathologic process is status spongiosus of gray matter [9, 46, 77]. Additionally, *Hogan* et al. [66] have recently demonstrated progressive degeneration of the rod outer and inner segments and photoreceptor nuclei of the retina in hamsters inoculated with the scrapie agent.

These agents are capable of evading the immune system because they are nonimmunogenic [46]. Their nonimmunogenicity can be inferred from the lack of an inflammatory response in the CNS, absence of pleocytosis in CSF, and from the fact that infected animals show no evidence of an immune response to the causative agent [46, 147]. Moreover, infected animals exhibit intact T and B lymphocyte functions in vivo and in vitro, and additionally, there are no alterations in the pathogenesis of slow virus diseases by immunosuppression or immunopotentiation [46].

Recently, *Sotelo* et al. [164] have demonstrated autoantibodies in patients with kuru and CJD, directed against axonal neurofilaments. These studies suggested an immune response associated with these infections, directed not against the agents themselves but rather against 'self-antigens'. The autoantibodies detected were not specific to CJD and kuru; sera from patients with other neurological disorders and normal donors also contained the autoantibody but at lower levels than sera from CJD and kuru patients. Clearly, characterization of the unconventional agents would be useful to elucidation of the pathogenesis of these diseases.

VI. Defective Interfering Particles and Persistent Infections

One mechanism which may be involved in the establishment and maintenance of persistence is the generation of defective interfering (DI) particles (fig. 2). A DI particle is operationally defined as a virus which is defective in that it cannot autonomously replicate unless the cell is co-infected with standard wild-type (wt) virus. DI particles also interfere with the normal replication of wt virus [70]. The general biology of DI particles has been reviewed by *Huang and Baltimore* [70].

DI particles may be involved in viral persistence by interacting with the host cell and wt virus, preventing normal wt lysis of the infected cell, thus allowing cell survival. Viral genetic information, albeit probably incomplete due to the defectiveness of the DI particle, may be preserved and persist in cells co-infected with wt virus and homologous DI particles.

DI particles have been generated in vitro by high multiplicity passage of viruses of a number of classes [70], and have been shown to play a role in establishment and maintenance of persistence in vitro [2, 67, 68, 83, 84, 139, 154, 156, 159]. Differences in the mechanisms of DI particle regulation in the establishment and maintenance of persistent viral infections in vitro exist among the various virus-cell interactions studied. Other viral particles which may be involved with persistence in vitro and in vivo are temperature-sensitive mutants [144].

Huang and Baltimore [70] first postulated that DI particles might play a role in viral persistence; DI particles have been used to establish persistent infections in vivo [70], but the role of DI particles in the maintenance of a natural persistent infection in vivo remains to be elucidated.

VII. The Role of Viral Persistence in Nature

In view of the fact that viruses of a wide number of classes cause periodic epi- or pandemics, it is germane to consider what effect viral persistence in a single animal may have on the population as a whole. A virus which persists in a single animal, without production of progeny, is caught in an evolutionary dead-end for the virus; death of the host will lead to cessation of viral spread throughout the population. Thus, persistence within an individual can only 'serve' the virus if the persistent infection allows for vertical or horizontal spread of the virus to another susceptible host. The persistently infected individual may provide a reservoir for viral genetic information between episodes of large-scale infections of the entire population. But a reservoir of viral genetic information must be expressed at least intermittently to spread the virus to other hosts. The examples of persistent, slow and latent viral infections presented above elaborate some of the mechanisms by which a virus may maintain itself within an individual, thereby providing it with a temporary 'escape' from herd immunity. Additionally, these examples suggest some mechanisms by which certain viruses may pass from one individual to another.

The frequency of induction of persistent infections during a naturally occurring epidemic is generally unknown. The paucity of data on persistent infections is not surprising considering that the persistent state does not always produce obvious clinical disease. Moreover, the prolonged period between primary exposure and eventual clinical disease renders the connection between the two difficult. It is clear that epidemiologic studies are required for each persistent virus-animal model to determine if the persistently infected carrier serves as a reservoir of viral genetic information, to what degree the information is expressed and under what conditions, and to what extent the virus is transmitted to other members of the population. The continued study of persistent viral infections is warranted because of the far-reaching effects a persistently infected host may have on a naive population.

References

1 Agnarsdotter, G.: Subacute sclerosing panencephalitis; in Waterson, Recent advances in clinical virology, pp. 21–49 (Churchill-Livingstone, London 1977).

2 Ahmed, R.; Graham, A.F.: Persistent infection in L cells with temperature sensitive mutants of reovirus. J. Virol. *23:* 250–262 (1977).

3 Albrecht, P.: Immune control in experimental subacute sclerosing panencephalitis. Am. J. clin. Path. *70:* 175–184 (1978).

4 Amerding, D.; Rossiter, H.: Induction of cytolytic T- and B-cell responses against influenza virus infections. Infect. Immunity *28:* 799–811 (1980).

5 Baglioni, C.: Interferon-induced enzymatic activities and their role in antiviral state. Cell *17:* 255–264 (1979).

6 Banks, K.L.; Henson, J.B.; McGuire, T.C.: Immunologically mediated glomerulonephritis of horses. I. Pathogenesis of persistent infection by equine infectious anemia virus. Lab. Invest. *26:* 701–707 (1972).

7 Baringer, J.R.: Recovery of herpes simplex virus from human sacral ganglions. New Engl. J. Med. *291:* 828–830 (1974).

8 Baringer, J.R.: Herpes simplex infection of nervous tissue in animals and man. Prog. med. Virol., vol. 20, pp. 1–26 (Karger, Basel 1975).

9 Baringer, J.R.: Virus diseases of the nervous system; in Tyler, Dawson, Current neurology, vol. 2 (Houghton-Mifflin, Boston 1979).

10 Baringer, J.R.; Griffith, J.F.: Experimental herpes simplex encephalitis: early neuropathologic changes. J. Neuropath. exp. Neurol. *29:* 89–104 (1970).

11 Baringer, J.R.; Swoveland, P.: Recovery of herpes simplex virus from human trigeminal ganglions. New Engl. J. Med. *288:* 648–650 (1973).

12 Baringer, J.R.; Swoveland, P.: Persistent herpes simplex virus infection in rabbit trigeminal ganglia. Lab. Invest. *30:* 230–240 (1974).

13 Bastian, F.O.; Rabson, A.S.; Yee, C.L.; Tralka, T.S.: Herpes virus hominis: isolation from human trigeminal ganglion. Science *178:* 306–307 (1972).

14 Bishop, J.M.: Retroviruses. A. Rev. Biochem. *47:* 35–88 (1978).

15 Bloom, M.E.; Race, R.E.; Wolfinbarger, J.B.: Characteristics of Aleutian disease as a parvovirus. J. Virol. *35:* 836–843 (1980).

16 Bodechtel, G.; Guttman, E.: Diffuse Encephalitis mit sklerosierender Entzündung des Hemisphärenmarkes. Z. ges. Neurol. Psychiat. *133:* 601–619 (1931).

17 Brinton-Darnell, M.; Collins, J.K.; Plageman, P.G.W.: Lactate dehydrogenase-elevating virus replication, maturation, and viral RNA synthesis in primary mouse macrophage cultures. Virology *65:* 187–195 (1975).

18 Byington, D.P.; Johnson, K.P.: Experimental subacute sclerosing panencephalitis in the hamster: correlation of age with chronic inclusion encephalitis. J. infect. Dis. *126:* 18–26 (1973).

19 Carrigan, D.R.; Johnson, K.P.: Chronic, relapsing myelitis in hamsters associated with experimental measles virus infection. Proc. natn. Acad. Sci. USA *77:* 4297–4300 (1980).

20 Carton, C.A.: Effect of previous sensory loss on the appearance of herpes simplex. J. Neurosurg. *10:* 463–468 (1953).

21 Carton, C.A.; Kilbourne, E.D.: Activation of latent herpes simplex by trigeminal sensory-root section. New Engl. J. Med. *246:* 172–176 (1952).

22 Cho, H.J.: Purification and structure of Aleutian disease virus; in Kimberlin, Slow virus diseases of animals and man, chapt. 8, pp. 159–174 (North-Holland, Amsterdam 1976).

23 Clements, J.E.; Pedersen, F.S.; Narayan, O.; Haseltine, W.A.: Genomic changes associated with antigenic variation of Visna virus during persistent infection. Proc. natn. Acad. Sci. USA *77:* 4454–4458 (1980).

24 Cole, G.A.; Nathanson, N.: Lymphocytic choriomeningitis: pathogenesis. Prog. med. Virol., vol. 18, pp. 94–115 (Karger, Basel 1975).

25 Cook, M.L.; Stevens, J.G.: Pathogenesis of herpetic neuritis and ganglionitis in mice: evidence for intra-axonal transport of infection. Infect. Immunity *7:* 272–288 (1973).

26 Connolly, J.: Subacute sclerosing panencephalitis. J. clin. Path. *25:* supp. 6, pp. 73–77 (1972).

27 Coyle, P.K.; Wolinsky, J.S.; Griffin, D.E.; Char, D.H.: Circulating immune complexes in progressive rubella panencephalitis contain anti-rubella IgG. Ann. Neurol. (in press).

28 Crawford, T.B.; Cheevers, W.P.; Klvjer-Anderson, P.; McGuire, T.C.: Equine infectious anemia: virion characteristics, virus-cell interaction and host responses; in Stevens, Todaro, Fox, Persistent Viruses, vol. XI. ICN-UCLA Symp. on Molecular and Cell Biology, pp. 727–749 (Academic Press, New York 1978).

29 Cremer, N.E.; Oshiro, L.S.; Weil, M.L.; Lennette, E.H.; Itabashi, H.H.; Carnay, L.: Isolation of rubella virus from brain in chronic progressive panencephalitis. J. gen. Virol. *29:* 143–153 (1975).

30 Dal Canto, M.C.; Lipton, H.L.: Primary demyelination in Theiler's virus infection: an ultrastructural study. Lab. Invest. *33:* 626–632 (1975).

31 Dal Canto, M.C.; Lipton, H.L.: Animal model of human disease: multiple sclerosis. Am. J. Path. *88:* 497–500 (1977).

32 Dal Canto, M.C.; Lipton, H.L.: Schwann cell remyelination and recurrent demyelination in the central nervous system of mice infected with attenuated Theiler's virus. Am. J. Path. *98:* 101–110 (1980).

33 Daniels, J.B.; Pappenheimer, A.M.; Richardson, S.: Observations on encephalomyelitis of mice (DA strain). J. exp. Med. *96:* 517–530 (1952).

34 Dawson, J.: Cellular inclusions in cerebral lesions of lethargic encephalitis. Am. J. Path. *9:* 7–16 (1933).

35 deBoer, G.F.; Terfstra, C.; Houwers, D.J.: Studies of Zwoegerziekte (maedi) in the Netherlands, a review. Bull. Off. int. Epizoot. *89:* 487–506 (1978).

36 Docherty, J.J.; Dochan, M.: The latent herpes simplex virus. Bac. Rev. *38:* 337–355 (1974).

37 Doherty, P.C.; Zinkernagel, R.M.: T-cell-mediated immunopathology in viral infections. Transplant. Rev. *19:* 89–120 (1974).

38 Doherty, P.C.; Zinkernagel, R.M.: Specific immune lysis of paramyxovirus-infected cells by H-2 compatible thymus derived lymphocytes. Immunology *31:* 27 (1976).

39 Dorries, K.; Johnson, R.T.; ter Meulen, V.: Detection of polyomavirus in PML-brain tissue by *in situ* hybridization. J. gen. Virol. *42:* 49–57 (1978).

40 Duffey, P.S.; Martinez, D.; Abrams, G.D.; Murphey, H.: Pathogenetic mechanisms in immune polioencephalomyelitis: induction of disease in immunosuppressed mice. J. Immun. *116:* 475–481 (1976).

41 Erb, P.; Feldman, M.: The role of macrophages in the generation of T-helper cells. II. The genetic control of the macrophage-T cell interaction for helper cell induction with soluble antigen. J. exp. Med. *142:* 460 (1975).

42 Evans, R.: Replication of Riley's plasma enzyme elevating virus *in vitro*. J. gen. Microbiol. *37:* vii (1964).

43 Fenner, F.; McAuslan, B.R.; Mims, C.A.; Sambrook, J.; White, D.O.: The biology of animal viruses; 2nd ed., pp. 479–543 (Academic Press, New York 1974).

44 Fong, B.; Scriba, M.: Use of [^{125}I] deoxycytidine to detect herpes simplex virus-specific thymidine kinase in tissues of latently infected guinea pigs. J. Virol. *34:* 644–649 (1980).

45 Frei, P.C.; Benacerraf, B.; Thorbecke, G.J.: Phagocytosis of the antigen, a crucial step in the induction of the primary immune response. Proc. natn. Acad. Sci. USA *53:* 20 (1965).

46 Gajdusek, C.D.: Unconventional viruses and the origin and disappearance of Kuru. Science *197:* 943–959 (1977).

47 Galloway, D.A.; Fenoglio, C.; Shevchuck, M.; McDougall, J.K.: Detection of herpes simplex RNA in human sensory ganglia. Virology *95:* 265–268 (1979).

48 Greenfield, J.: Encephalitis and encephalomyelitis in England and Wales during the last decade. Brain *73:* 141–166 (1950).

49 Griffin, D.E.; Narayan, O.; Adams, R.J.: Early immune responses in Visna, a slow virus disease of sheep. J. infect. Dis. *138:* 340–350 (1978).

50 Haase, A.T.; Varmus, H.E.: Demonstration of a DNA provirus in the lytic growth of Visna virus. Nature new Biol. *245:* 237–239 (1973).

51 Haase, A.T.; Stowring, L.; Narayan, O.; Griffin, D.; Price, D.: Slow persistent infection caused by Visna virus: role of host restriction. Science *195:* 175–177 (1977).

52 Haase, A.T.; Brahic, M.; Carrol, D.; Scott, J.; Stowring, L.; Traynor, B.; Ventura, P.; Narayan, O.: Visna: an animal model for studies of virus persistence; in Stevens, Todaro, Fox, Persistent viruses, vol. XI. ICN-UCLA Symp. on Molecular and Cell Biology, pp. 643–661 (Academic Press, New York 1978).

53 Habicht, G.S.; Chiller, J.M.; Weigle, W.O.: Termination of acquired and natural immunologic tolerance with specific complexes. J. exp. Med. *142:* 312–316 (1975).

54 Hall, W.M.; Choppin, P.W.: Evidence for lack of synthesis of the M polypeptide of measles virus in brain cells in subacute sclerosing panencephalitis. Virology *99:* 443–447 (1979).

55 Harter, D.H.; Axel, R.; Burney, A.; Gulati, S.; Schlom, J.; Spiegelman, S.: The relationship of visna, maedi, and RNA tumor viruses as studied by molecular hybridization. Virology *52:* 287–291 (1973).

56 Henle, G.; Henle, W.; Diehl, V.: Relation of Burkitt's tumor-associated herpes-type virus to infectious mononucleosis. Proc. natn. Acad. Sci. USA *59:* 94–101 (1968).

57 Henle, G.; Henle, W.; Clifford, P.; Diehel, V.; Kafuko, G.W.; Kurya, B.G.; Klein, G.; Morrow, G.; Munube, G.M.R.; Pike, D.; Tukei, P.M.; Ziegler, J.L.: Antibodies to Epstein-Barr virus in Burkitt's lymphoma and control groups. J. natn. Cancer Inst. *43:* 1147–1157 (1969).

58 Henson, J.B.; Gorham, J.R.; Tanaka, Y.; Padgett, G.A.: The specific development of ultrastructural lesions in the glomeruli of mink with experimental Aleutian disease. Lab. Invest. *19:* 153–162 (1968).

59 Henson, J.B.; Gorham, J.R.; Padgett, G.A.; Davis, W.C.: Pathogenesis of the glomerular lesions in Aleutian disease of mink. Archs. Path. *87:* 21–28 (1969).

60 Henson, J.B.; McGuire, T.C.: Immunopathology of equine infectious anemia. Am. J. clin. Path. *56:* 306–314 (1971).

61 Henson, J.B.; McGuire, T.C.: Equine infectious anemia. Prog. med. Virol., vol. 18, pp. 143–149 (Karger, Basel 1974).

62 Henson, J.B.; Gorham, J.R.; McGuire, T.C.; Crawford, T.B.: Pathology and pathogenesis of Aleutian disease; in Kimberlin, Slow virus diseases of animals and man, chapt. 9, pp. 173–205 (North-Holland, Amsterdam 1976).

63 Hill, T.J.; Field, H.J.; Blyth, W.A.: Acute and recurrent infection with herpes simplex virus in the mouse: a model for studying latency and recurrent disease. J. gen. Virol. *28:* 341–353 (1975).

64 Hill, T.J.; Blyth, W.A.; Harbour, D.A.: Trauma to the skin causes recurrence of herpes simplex in the mouse. J. gen. Virol. *39:* 21–28 (1978).

65 Ho, M.: Role of specific cytotoxic lymphocytes in cellular immunity against murine cytomegalovirus. Infect. Immunity *27:* 767–776 (1980).

66 Hogan, R.N.; Baringer, J.R.; Prusiner, S.B.: Progressive retinal degeneration in scrapie-infected hamsters. Lab. Invest. *44:* 34–42 (1981).

67 Holland, J.J.; Villareal, L.P.: Persistent noncytocidal vesicular stomatitis virus infections mediated by defective T particles that suppress virion transcriptase. Proc. natn. Acad. Sci. USA *71:* 2956–2960 (1974).

68 Holland, J.J.; Villareal, L.P.; Welsh, R.M.; Oldstone, M.B.A.; Kohne, D.; Lazzarini, R.; Scolnick, E.: Long-term persistent vesicular stomatitis virus and rabies virus infections of cells *in vitro.* J. gen. Virol. *233:* 193–211 (1976).

69 Huang, A.S.; Baltimore, D.: Defective viral particles and viral disease processes. Nature, Lond. *226:* 325–327 (1970).

70 Huang, A.S.; Baltimore, D.: Defective interfering animal viruses; in Fraenkel-Conrat, Wagner, Comprehensive virology, vol. 10, chapt. 2, pp. 73–116 (Plenum Press, New York 1977).

71 Ingram, D.G.; Cho, H.J.: Aleutian disease in mink: virology, immunology, and pathogenesis. J. Rheumatol. *1:* 74–92 (1974).

72 Jan, J.E.; Tingle, A.J.; Donald, G.; Kettyls, M.; Buckler, W.S.J.; Dolman, C.L.: Progressive rubella panencephalitis: clinical course and response to 'isoprinosine'. Devl Med. Child Neur. *21:* 648–652 (1979).

73 Johannes, R.; Sever, J.: Subacute sclerosing panencephalitis. A. Rev. Med. *26:* 589–601 (1975).

74 Johnson, K.; Byington, D.P.; Gaddis, L.: Subacute sclerosing panencephalitis. Adv. Neurol. *6:* 77–86 (1974).

75 Johnson, K.P.; Norrby, E.: Subacute sclerosing panencephalitis (SSPE) agent in hamsters. III. Induction of defective measles infection in hamster brain. Exp. molec. Path. *21:* 116–172 (1974).

76 Johnson, K.P.; Wolinsky, J.S.; Ginsberg, A.H.: Immune mediated syndromes of the nervous system related to virus infections; in Klawans, Vinken, Bruyn, Infections of the nervous system. Handbook of clinical neurology, vol. 34, chapt. 20, pp. 391–434 (North-Holland, Amsterdam 1978).

77 Johnson, K.P.; Wolinsky, J.S.; Swoveland, P.: Central nervous system infections (chronic); in Hsiung, Green, CRC handbook series in clinical laboratory science, vol. 1, pp. 295–307 (CRC Press, Palm Beach 1978).

78 Johnson, K.P.; Norrby, E.; Swoveland, P.; Carrigan, D.R.: Experimental subacute sclerosing panencephalitis: measles matrix (M) protein selectively disappears during subacute CNS infection. J. infect. Dis. *144:* 161–169 (1981).

79 Johnson, R.T.: The pathogenesis of herpes virus encephalitis. I. Virus pathways to the nervous system of suckling mice demonstrated by fluorescent antibody staining. J. exp. Med. *119:* 343–356 (1964).

80 Johnson, R.T.; Narayan, O.; Weiner, L.P.; Greenlee, J.E.: Progressive multifocal leukoencephalopathy; in ter Meulen, Katz, Slow virus infection of the central nervous system, pp. 91–100 (Plenum Press, New York 1977).

81 Kappler, J.W.; Marrack, P.C.: Helper T cells recognize antigen and macrophage components simultaneously. Nature, Lond. *262:* 797 (1976).

82 Karstad, L.; Pridham, T.J.: Aleutian disease of mink. I. Evidence of its viral etiology. Can. J. comp. Med. *26:* 97–102 (1962).

83 Kawai, A.; Matsumoto, S.: Interfering and noninterfering defective particles generated by a rabies small plaque variant virus. Virology *76:* 60–71 (1977).

84 Kawai, A.; Matsumoto, S.; Tanabe, K.: Characterization of rabies viruses recovered from persistently infected BHK cells. Virology *67:* 520–533 (1975).

85 Klein, R.J.: Pathogenic mechanisms of recurrent herpes simplex virus infections. Archs. Virol. *51:* 1–13 (1976).

86 Knotts, F.B.; Cook, M.L.; Stevens, J.G.: Latent herpesvirus in the central nervous system of rabbits and mice. J. exp. Med. *138:* 740–744 (1973).

87 Kobayashi, K.; Henson, J.B.; Gorham, J.R.: Viral neutralizing antibodies in equine infectious anemia. Fed. Proc. *28:* 429 (1969).

88 Kono, Y.; Kobayashi, K.; Fukunaga, Y.: Antigenic drift of equine infectious anemia virus in chronically infected horses. Arch. ges. Virusforsch. *41:* 1–10 (1973).

89 Kosnowski, U.; Ertl, H.: Lysis mediated by T cells and restricted by H-2 antigen of target cells infected with vaccinia virus. Nature, Lond. *255:* 522 (1975).

90 Levine, M.; Goldin, A.L.; Glorioso, J.C.: Persistence of herpes simplex virus genes in cells of neuronal origin. J. Virol. *35:* 203–210 (1980).

91 Levy, J.A.: C-type RNA viruses and autoimmune disease; in Talal, Autoimmunity – genetic, immunologic, virologic, and clinical aspects, chapt. 12 (Academic Press, New York 1977).

92 Lipton, H.L.: Theiler's virus infection in mice: an unusual biphasic disease process leading to demyelination. Infect. Immunity *11:* 1141–1155 (1975).

93 Lipton, H.L.: Characterization of the T0 strains of Theiler's mouse encephalomyelitis viruses. Infect. Immunity *20:* 869–872 (1978).

94 Lipton, H.L.: Persistent Theiler's murine encephalomyelitis virus infection in mice depends on plaque size. J. gen. Virol. *46:* 169–177 (1980).

95 Lipton, H.L.; Dal Canto, M.C.: Susceptibility of inbred mice to chronic central nervous system infection by Theiler's murine encephalomyelitis virus. Infect. Immunity *26:* 369–374 (1979).

96 Lipton, H.L.; Dal Canto, M.C.: Chronic neurologic disease in Theiler's virus infection of SJL/J mice. J. neurol. Sci. *30:* 201–207 (1976).

97 Lipton, H.L.; Dal Canto, M.C.: Theiler's virus-induced demyelination: prevention by immunosuppression. Science *192:* 62–64 (1976).

98 Lipton, H.L.; Dal Canto, M.C.: Contrasting effects of immunosuppression on Theiler's virus infection in mice. Infect. Immunity *15:* 903–909 (1977).

99 Lipton, H.L.; Dal Canto, M.C.: The T0 strains of Theiler's viruses cause a 'slow virus-like' infection in mice. Ann. Neurol. *6:* 25–28 (1979).

100 Lipton, H.L.; Friedman, A.: Purification of Theiler's murine encephalomyelitis virus and analysis of the structural virion polypeptides: correlation of the polypeptide profile with virulence. J. Virol. *33:* 1165–1172 (1980).

101 Liu, C.; Collins, J.; Sharp, E.: The pathogenesis of Theiler's GDVII encephalomyelitis virus infection in mice as studied by immunofluorescent techniques and infectivity titrations. J. Immun. *98:* 46–55 (1967).

102 Lundquist, R.E.; Sullivan, M.; Maizel, J.V.: Characterization of a new isolate of poliovirus defective interfering particles. Cell *18:* 759–769 (1979).

103 Martin, J.R.; Nathanson, N.: Animal models of virus-induced demyelination. Prog. Neuropathol. *4:* 27–50 (1979).

104 Martinez, D.; Brinton, M.; Tachovsky, T.J.; Phelps, A.H.: Identification of lactate dehydrogenase-elevating virus as the etiologic agent of genetically restricted, age-dependent polioencephalomyelitis of mice. Infect. Immunity *27:* 979–987 (1980).

105 Menser, M.A.; Dodds, L.; Harley, J.D.: A 25 year follow-up of congenital rubella. Lancet *ii:* 1347–1350 (1967).

106 Miller, G.; Niedermann, J.C.; Andrews, L.: Prolonged oropharyngeal excretion of Epstein-Barr virus after infectious mononucleosis. New Engl. J. Med. *288:* 229–232 (1973).

107 Miller, G.; Shope, T.; Lisco, H.; Stitt, D.; Lipman, H.: Epstein-Barr virus: transformation, cytopathic changes and viral antigens in squirrel monkey and marmoset leukocytes. Proc. natn. Acad. Sci. USA *69:* 383–387 (1972).

108 Mims, C.A.: Factors in the mechanism of persistence of viral infections. Prog. med. Virol., vol. 18, pp. 1–14 (Karger, Basel 1974).

109 Morrison, W.I.; Wright, N.G.: Viruses associated with renal disease of man and animals. Prog. med. Virol., vol. 23, pp. 22–50 (Karger, Basel 1977).

110 Murphy, W.H.; Tam, M.R.; Lanzi, R.L.; Abell, A.R.; Kaufman, C.: Age dependence of immunologically induced central nervous system disease in C58 mice. Cancer Res. *30:* 1612–1622 (1970).

111 McFarland, H.F.: *In vitro* studies of cell-mediated immunity in acute viral infections. J. Immun. *113:* 173 (1974).

112 McGuire, T.C.; Henson, J.B.; Burger, D.: Complement (C'3)-coated red blood cells following infections with the virus of equine infectious anemia. J. Immun. *103:* 293 (1969).

113 McGuire, T.C.; Henson, J.B.; Quist, S.E.: Viral induced hemolysis in equine infectious anemia. Am. J. vet. Res. *30:* 2091 (1969).

114 McGuire, T.C.; Crawford, T.B.; Henson, J.B.: Immunofluorescent localization of equine infectious anemia virus in tissue. Am. J. Path. *62:* 283 (1971).

115 McGuire, T.C.; van Hoosier, G.L.; Henson, J.B.: The complement fixation reaction in equine infectious anemia. Demonstration of inhibition by IgG(T). J. Immun. *107:* 1738–1744 (1971).

116 McGuire, T.C.; Crawford, T.B.; Henson, J.B.: Equine infectious anemia. Demonstration of infectious virus-antibody complexes in the serum. Immunol. Commun. *1:* 545–551 (1972).

117 Narayan, O.; Griffin, D.E.; Chase, J.: Antigenic shift of visna virus in persistently infected sheep. Science *199:* 376–378 (1977).

118 Narayan, O.; Griffin, D.E.; Clements, J.E.: Virus mutation during slow infection. Temporal development and characterization of mutants of visna virus recovered from sheep. J. gen. Virol. *41:* 343–352 (1978).

119 Narayan, O.; Griffin, D.E.; Clements, J.E.: Progressive antigenic drift of visna virus in persistently infected sheep; in Stevens, Todaro, Fox, Persistent viruses. ICN-UCLA Symp. on Molecular and Cell Biology, vol. XI, pp. 663–669 (Academic Press, New York 1978).

120 Notkins, A.L.: Lactic dehydrogenase virus. Bact. Rev. *29:* 143–160 (1965).

121 Notkins, A.L.: Effect of virus infections on immune function; in Miescher, 6th Int. Symp. Immunopathol., pp. 413–425 (Springer, Berlin 1970).

122 Notkins, A.L.: Enzymatic and immunologic alterations in mice infected with lactic dehydrogenase virus. Am. J. Path. *64:* 733–746 (1971).

123 Notkins, A.L.; Mahar, S.; Scheele, C.; Goffman, J.: Infectious virus-antibody complex in the blood of chronically infected mice. J. exp. Med. *124:* 81–97 (1966).

124 Notkins, A.L.; Sheele, C.: Impaired clearance of enzymes in mice infected with the lactic dehydrogenase agent. J. natn. Cancer Inst. *33:* 741–749 (1964).

125 Oldstone, M.B.A.; Dixon, F.J.: Lactic dehydrogenase virus-induced immune complex type of glomerulonephritis. J. Immun. *106:* 1260–1263 (1971).

126 Olitsky, P.K.: A transmissible agent (Theiler's virus) in the intestines of normal mice. J. exp. Med. *72:* 113–127 (1940).

127 Openshaw, H.; Shavrina-Asher, L.V.; Wohlenberg, C.; Sekizawa, T.; Notkins, A.L.: Acute and latent infection of sensory ganglia with herpes simplex virus: immune control and virus reactivation. J. gen. Virol. *44:* 205–215 (1979).

128 Padgett, B.L.; Walker, D.L.; Zu Rhein, G.M.; Eckroode, R.J.; Dessel, B.H.: Cultivation of papova-like virus from human brain with progressive multifocal leukoencephalopathy. Lancet *i:* 1257–1260 (1971).

129 Padgett, B.L.; Walker, D.L.: New human papovavirus. Prog. med. Virol., vol. 22, pp. 1–35 (Karger, Basel 1976).

130 Panitch, H.S.; Hooper, C.J.; Johnson, K.P.: CSF antibody to myelin basic protein: measurement in patients with multiple sclerosis and subacute sclerosing panencephalitis. Archs Neurol., Chicago *37:* 206–209 (1980).

131 Payne, F.; Baublis, J.V.: Measles virus and subacute sclerosing panencephalitis; in Pollard, Perspectives in virology, vol. VII, pp. 179–185 (Academic Press, New York 1971).

132 Penney, J.B.; Wolinsky, J.S.: Neuronal and oligodendroglial infection by the WW strain of Theiler's virus. Lab. Invest. *40:* 324–330 (1979).

133 Pette, H.; Doring, G.: Über eine einheimische Panencephalomyelitis vom Charakter der Encephalitis japonica. Dt. Z. Nervheilk. *149:* 7–44 (1939).

134 Petursson, G.; Nathanson, N.; Georgsson, G.; Panitch, H.; Palsson, P.A.: Pathogenesis of visna. I. Sequential virologic, serologic and pathologic studies. Lab. Invest. *35:* 402–412 (1976).

135 Pierce, C.W.; Kapp, J.A.; Benacerraf, B.: Regulation by the H-2 gene complex of macrophage-lymphoid cell interactions in secondary antibody response *in vitro*. J. exp. Med. *114:* 371 (1976).

136 Plagemann, P.G.W.; Gregory, K.F.; Swim, H.E.; Chan, K.K.W.: Plasma dehydrogenase elevating agent of mice: distribution in tissues and effect on lactic dehydrogenase isozyme patterns. Can J. Microbiol. *9:* 75–86 (1963).

137 Plagemann, P.G.W.; Swim, H.E.: Propagation of lactic dehydrogenase-elevating virus in cell culture. Proc. Soc. exp. Biol. Med. *121:* 1147–1152 (1966).

138 Pope, J.H.; Rowe, W.P.: Identification of WM 1 as LDH virus, and its recovery from wild mice in Maryland. Proc. Soc. exp. Biol. Med. *116:* 1015–1019 (1964).

139 Popescu, M.; Lehmann-Grube, F.: Defective interfering particles in mice infected with lymphocytic choriomeningitis virus. Virology *77:* 78–83 (1977).

140 Porter, D.D.; Dixon, F.J.; Larsen, A.E.: Metabolism and function of gamma globulin in Aleutian disease of mink. J. exp. Med. *121:* 889–900 (1965).

141 Porter, D.D.; Larsen, A.E.; Porter, H.G.: The pathogenesis of Aleutian disease of mink. I. *In vivo* viral replication and the host antibody response to viral antigen. J. exp. Med. *130:* 575–589 (1969).

142 Porter, D.D.; Larsen, A.E.; Porter, H.G.: The pathogenesis of Aleutian disease of mink. III. Immune complex arteritis. Am. J. Path. *71:* 331–338 (1973).

143 Porter, D.D.; Porter, H.G.: Deposition of immune complexes in the kidneys of mice infected with lactic dehydrogenase virus. J. Immun. *106:* 1264–1266 (1971).

144 Preble, O.T.; Youngner, J.S.: Temperature sensitive viruses and the etiology of chronic and inapparent infections. J. infect. Dis. *131:* 467–473 (1975).

145 Price, R.W.; Walz, M.A.; Wohlenberg, C.; Notkins, A.L.: Latent infection of sensory ganglia with herpes simplex virus; efficacy of immunization. Science *188:* 938–940 (1975).

146 Prusiner, S.B.; Hadlow, W.J.: Slow transmissible diseases of the nervous system, vol. 1 (Academic Press, New York 1979).

147 Prusiner, S.B.; Hadlow, W.J.: Slow transmissible diseases of the nervous system, vol. 2 (Academic Press, New York 1979).

148 Puga, A.; Rosenthal, J.D.; Openshaw, H.; Notkins, A.L.: Herpes simplex virus DNA and mRNA sequences in acutely and chronically infected trigeminal ganglia of mice. Virology *89:* 102–111 (1978).

149 Rajcani, J.; Matis, J.; Field, H.: Immediate early and early herpesvirus antigens were not seen in serial sections of ganglia with established latent infection. Abstr. Cold Spring Harb. Meet. Herpesviruses, p. 121 (1979).

150 Riley, V.; Lilly, F.; Huerto, E.; Bardell, D.: Transmissible agent associated with 26 types of experimental mouse neoplasms. Science *132:* 545–547 (1960).

151 Roizman, B.: An inquiry into the mechanism of recurrent herpes infections of man; in Pollard, Perspectives in virology, vol. 4, pp. 283–301 (Harper & Row, New York 1965).

152 Rosenthal, H.S.; Shevach, E.M.: Function of macrophages in antigen recognition by guinea pig T-lymphocytes. I. Requirement for histocompatible macrophages and lymphocytes. J. exp. Med. *138:* 1194–1212 (1973).

153 Rosenthal, A.S.; Barcinski, A.; Blake, T.J.: Determinant selection is a macrophage dependent immune response gene function. Nature, Lond. *267:* 156 (1977).

154 Rowlands, D.; Grabau, E.; Spindler, K.; Jones, C.; Semler, B.; Holland, J.: Virus protein changes and RNA terminal alterations evolving during persistent infections. Cell *19:* 871–880 (1980).

155 Schilder, P.: Die encephalitis periaxialis diffusa. Arch. Psychiat. NervKrankh. *71:* 327–356 (1924).

156 Schmaljohn, C.; Blair, C.D.: Persistent infection of cultured mammalian cells by Japanese encephalitis virus. J. Virol. *24:* 580–589 (1977).

157 Schuller, E.; Delasnerie, N.; Allinquant, B.: Intrathecal rubella and RNA antibody synthesis in multiple sclerosis and progressive rubella panencephalitis. Biomedicine *27:* 139–141 (1977).

158 Scott, J.V.; Stowring, L.; Haase, A.T.; Narayan, O.; Vigne, R.: Antigenic variation in Visna virus. Cell *18:* 321–327 (1979).

159 Semler, B.L.; Holland, J.: Persistent vesicular stomatitis virus infection mediates base substitutions in viral RNA termini. J. Virol. *32:* 420–428 (1979).

160 Schevach, E.M.; Rosenthal, A.S.: Function of macrophages in antigen recognition by guinea pig T-lymphocytes. II. Role of the macrophage in the regulation of genetic control of the immune response. J. exp. Med. *138:* 1213–1229 (1973).

161 Sigurdsson, B.: Observation of three slow virus infections of sheep. Maedi, paratuberculosis, Rida, a chronic encephalitis of sheep, with general remarks on infections which develop slowly, and some of their special characteristics. Br. vet. J. *110:* 225–270, 307–322, 341–354 (1954).

162 Snodgrass, M.J.; Lowrey, D.S.; Hanna, M.G.: Changes induced by lactic dehydrogenase virus in thymus and thymus-dependent areas of lymphatic tissue. J. Immun. *108:* 877–892 (1972).

163 Sokawa, Y.; Ando, T.; Ishihara, Y.: Induction of 2′,5′-oligoadenylate synthetase and interferon in mouse trigeminal ganglia infected with herpes simplex virus. Infect. Immunity *28:* 719–723 (1980).

164 Sotelo, J.; Gibbs, C.J.; Gajdusek, D.C.: Autoantibodies against axonal neurofilaments in patients with Kuru and Creutzfeldt-Jakob disease. Science *210:* 190–193 (1980).

165 Stevens, J.G.: Latent herpes simplex virus and the nervous system. Curr. Top. Microbiol. Immunol. *70:* 31–50 (1975).

166 Stevens, J.G.; Cook, M.L.: Latent herpes simplex virus in spinal ganglia of mice. Science *173:* 843–845 (1971).

167 Stevens, J.G.; Cook, M.L.: Latent herpes simplex virus in sensory ganglia; in Pollard, Perspectives in virology, vol. VIII, pp. 171–188 (Academic Press, New York 1973).

168 Stevens, J.G.; Nesburn, A.B.; Cook, M.L.: Latent herpes simplex virus from trigeminal ganglia of rabbits with recurrent eye infection. Nature new Biol. *235:* 216–217 (1972).

169 Stevens, J.G.; Cook, M.L.; Jordan, M.C.: Reactivation of latent herpes simplex virus after pneumococcal pneumonia in mice. Infect. Immunity *11:* 635–639 (1975).

170 Stowring, L.; Haase, A.T.; Charman, H.P.: Serological definition of the lentivirus group of retroviruses. J. Virol. *29:* 523–528 (1979).

171 Stroop, W.G.; Baringer, J.R.: The biochemistry of Theiler's murine encephalomyelitis virus (WW strain) isolated from acutely infected mouse brain: Identification of a previously unreported polypeptide. Infect. Immun. *32*: 769–777 (1981).

172 Tenser, R.B.; Dunstan, M.E.: Herpes simplex virus thymidine kinase expression in infection of the trigeminal ganglion. Virology *99:* 417–422 (1979).

173 Tenser, R.B.; Miller, R.L.; Rapp, F.: Trigeminal ganglion infection by thymidine kinase-negative mutants of herpes simplex virus. Science *205:* 915–917 (1979).
174 ter Meulen, V.: SSPE and measles virus: current state of our knowledge. Neuropädiatrie *4:* 347–349 (1973).
175 ter Meulen, V.; Hall, W.M.: Slow virus infections of the nervous system: virological, immunological and pathogenetic considerations. J. gen. Virol. *41:* 1–25 (1978).
176 ter Meulen, V.; Katz, M.; Muller, D.: Subacute sclerosing panencephalitis: a review. Curr. Trends Microbiol. *57:* 1–38 (1972).
177 Theiler, M.: Spontaneous encephalomyelitis of mice – a new disease. Science *80:* 122–123 (1934).
178 Theiler, M.: Spontaneous encephalomyelitis of mice – a new disease. J. exp. Med. *65:* 705–719 (1937).
179 Theiler, M.; Gard, S.: Encephalomyelitis of mice. I. Characteristics and pathogenesis of the virus. J. exp. Med. *72:* 49–67 (1940).
180 Theiler, M.; Gard, S.: Encephalomyelitis of mice. III. Epidemiology. J. exp. Med. *72:* 79–90 (1940).
181 Thormar, H.; Wisniewski, H.M.; Lin, F.H.: Sera and cerebrospinal fluids from normal and uninfected sheep contain a visna virus inhibiting factor. Nature, Lond. *279:* 245–246 (1979).
182 Townsend, J.J.; Baringer, J.R.; Wolinsky, J.S.; Malamud, N.; Mednick, J.P.; Panitch, H.S.; Scott, R.; Oshiro, L.S.; Cremer, N.E.: Progressive rubella panencephalitis; late onset after congenital rubella. New Engl. J. Med. *292:* 990–993 (1975).
183 Townsend, J.J.; Wolinsky, J.S.; Baringer, J.R.: The neuropathology of progressive rubella panencephalitis of late onset. Brain *99:* 81–90 (1976).
184 Townsend, J.J.; Stroop, W.G.; Baringer, J.R.; Wolinsky, J.S.; McKerrow, J.H.; Berg, B.O.: Neuropathology of progressive rubella panencephalitis after childhood rubella. Neurology (in press).
185 Turner, W.; Ebert, P.S.; Bassin, R.; Spahn, G.; Chirigos, M.A.: Potentiation of murine sarcoma viruses (Harvey) (Moloney) oncogenicity in lactic dehydrogenase-elevating virus-infected mice. Proc. Soc. exp. Biol. Med. *136:* 1314–1318 (1971).
186 Unanue, E.R.; Askonas, B.A.: Persistence of immunogenicity of antigen after uptake by macrophages. J. exp. Med. *127:* 915–926 (1968).
187 Unanue, E.R.; Cerottini, J.C.: The immunogenicity of antigen bound to the plasma membrane of macrophages. J. exp. Med. *131:* 717–725 (1970).
188 van Bogaert, L.: Une leucoencéphalite sclérosante subaigue. J. Neurol. Neurosurg. Psychiat. *8:* 101–120 (1945).
189 Vogt, P.K.: Genetics of RNA tumor viruses; in Fraenkel-Conrat, Wagner, Comprehensive virology, vol. 9, chapt. 9, pp. 341–455 (Plenum Press, New York 1977).
190 Walz, M.A.; Yamamoto, H.; Notkins, A.L.: Immunological response restricts number of cells in sensory ganglia infected with herpes simplex virus. Nature, Lond. *264:* 554–556 (1976).
191 Weigle, W.O.: Cellular events in experimental autoimmune thyroiditis, allergic encephalomyelitis and tolerance to self; in Talal, Autoimmunity – genetic, immunologic, virologic, and clinical aspects, pp. 141–170 (Academic Press, New York 1977).
192 Weil, M.L.; Itabashi, H.H.; Cremer, N.E.: Chronic progressive panencephalitis due to rubella virus simulating subacute sclerosing panencephalitis. New Engl. J. Med. *292:* 994–998 (1975).

193 Weinberg, R.A.: Integrated genomes of animal viruses. A. Rev. Biochem. *49:* 197–226 (1980).

194 Weiner, L.P.; Herndon, R.M.; Narayan, O.; Johnson, R.T.: Further studies of a simian virus 40-like virus isolated from human brain. J. Virol. *10:* 147–149 (1972).

195 Weiner, L.P.; Herndon, R.M.; Narayan, O.; Johnson, R.T.; Shah, K.; Rubinstein, C.J.; Prezioski, T.J.; Cowley, F.R.: Isolation of virus related to SV40 from patients with progressive multifocal leukoencephalopathy. New Engl. J. Med. *286:* 385–390 (1972).

196 Wild, T.F.; Giraudon, P.; Bernard, A.; Huppert, J.: Isolation and characterization of a defective measles virus from a subacute sclerosing panencephalitis patient. J. med. Virol. *4:* 103–114 (1979).

197 Wolinsky, J.S.: Progressive rubella panencephalitis; in Vinken, Bruyn, Handbook of clinical neurology, vol. 34 (North-Holland, New York 1978).

198 Wolinsky, J.S.; Berg, B.O.; Maitland, C.J.: Progressive rubella panencephalitis, case report. Archs. Neurol., Chicago *33:* 722–723 (1976).

199 Wolinsky, J.S.; Dau, P.C.; Buimovici-Klein, E.; Mednick, J.; Berg, B.O.; Lang, Z.B.; Coopers, L.Z.: Progressive rubella panencephalitis: immunovirological studies and results of isoprinosine therapy. Clin. exp. Immunol. *35:* 397–404 (1979).

200 Wong, C.Y.; Woodruff, J.J.; Woodruff, J.F.: Generation of cytotoxic T lymphocytes during coxsackie B-3 infection. I. Model and viral specificity. J. Immun. *118:* 1159–1164 (1977).

201 Wong, C.Y.; Woodruff, J.J.; Woodruff, J.F.: Generation of cytotoxic T lymphocytes during coxsackie B-3 infection. II. Characterization of effector cells and demonstration of cytotoxicity against viral-infected myofibers. J. Immun. *118:* 1165–1169 (1977).

202 Wroblewska, Z.; Gilden, D.; Wellish, M.; Rorke, L.B.; Warren, K.G.; Wolinsky, J.S.: Virus-specific intracytoplasmic inclusions in mouse brain produced by a newly isolated strain of Theiler virus. I. Virologic and morphologic studies. Lab. Invest. *37:* 595–602 (1977).

203 Wroblewska, Z.; Kim, S.U.; Sheffield, W.D.; Gilden, D.H.: Growth of the WW strain of Theiler virus in mouse central nervous organotypic cultures. Acta neuropath. *47:* 13–19 (1979).

204 Yaffe, D.: The distribution and *in vitro* propagation of an agent causing high plasma dehydrogenase activity. Cancer Res. *22:* 573–580 (1962).

205 Zinkernagel, R.M.: H-2 restriction of cell-mediated virus-specific immunity and immunopathology: self-recognition, altered self, and autoaggression; in Talal, Autoimmunity – genetic, immunologic, virologic, and clinical aspects, pp. 363–384 (Academic Press, New York 1977).

206 Zinkernagel, R.M.; Doherty, P.C.: Restriction of in vitro T cell-mediated cytotoxicity in lymphocytic choriomeningitis with a syngeneic or allogeneic system. Nature, Lond. *248:* 701–702 (1974).

207 Zinkernagel, R.M.; Doherty, P.C.: Immunological surveillance against altered self components by sensitized T lymphocytes in lymphocytic choriomeningitis. Nature, Lond. *251:* 547–548 (1974).

208 Zinkernagel, R.M.; Doherty, P.C.: Characteristics of the interaction *in vitro* between thymus-derived lymphocytes and target monolayers infected with lymphocytic choriomeningitis virus. Scand. J. Immunol. *3:* 287 (1974).

209 Zinkernagel, R.M.; Doherty, P.C.: H-2 compatibility requirement of T cell lysis of cells infected with LCM. Different cytotoxic T cell specificities are associated with structures coded in H-2K or H-2D. J. exp. Med. *141:* 1427–1436 (1975).

210 Zinkernagel, R.M.; Oldstone, M.B.A.: Cells that express viral antigens but lack H-2 determinants are not lysed by immune thymus-derived lymphocytes but are lysed by other antiviral immune attack mechanisms. Proc. natn. Acad. Sci. USA *73:* 3666–3670 (1976).

William G. Stroop, PhD, The Wistar Institute, 36th and Spruce Streets, Philadelphia, PA 19104 (USA)

Prog. med Virol., vol. 28, pp. 44–64 (Karger, Basel 1982)

Cytomegalovirus Infection in Transplant Patients

Robert F. Betts

Infectious Diseases Unit, Department of Medicine, University of Rochester School of Medicine, Rochester, N.Y., USA

Contents

I. Introduction

Very soon after the first renal allotransplantation had been undertaken it became apparent that infection with cytomegalovirus (CMV) contributed to illness in transplant patients [*Rifkind* et al., 1964; *Hedley-Whyte and Craighead,* 1965]. The initial data that implicated CMV as a cause of transplant-associated illness were based on histologic evidence of CMV often accompanying histologic evidence for other opportunistic organisms. Over the next several years – as it became clear, using virus isolation as a measure, how very common CMV infection was and how frequently it was recovered from asymptomatic subjects – there was doubt as to whether it played an important role in problems occurring in transplant patients. At one point, CMV was considered a harmless concomitant of the transplant process. However, multiple studies have now defined the very important role it plays in the transplant population. This review will attempt to explore these studies, highlighting early observations which helped develop an understanding, and then detail the more recent observations which have brought a greater understanding of the scope of the problem with CMV infection in transplantation.

II. Epidemiology of CMV Infection

Virus infection in transplant patients is of two types, reactivation of latent virus and development of primary virus infection. Those who are likely to develop reactivation of latent infection can be predicted on the basis of epidemiologic considerations. There is substantial information to suggest that the presence of antibody in a population indicates the presence of latent infection whereas the absence of antibody similarly indicates the absence of latent infection [*Betts* et al., 1975; *Yeager* et al., 1981]. In the majority of populations studied, the age at which CMV antibody is acquired is dependent on race, socioeconomic conditions, and sex [*Li and Hanshaw,* 1967; *Jordan* et al., 1973; *Luby and Shasby,* 1972]. In the poor, and even more so in the poor black population, frequency of antibody in young subjects (10 years old and younger) is very high, approaching 80%. The explanation for this high frequency of infection may be related to the common occurrence of CMV in cervical secretions and/or breast milk. About 15% of antibody-positive 15-year-old women shed virus in their cervix [*Knox* et al., 1979], and between 15 and 50% of lactating women shed virus in breast milk [*Hayes* et al., 1972; *Stagno* et al., 1980]. With increasing age there is a decreasing frequency of women who shed virus in the cervix; approximately 10% by age 25 and 0% by age 30 years [*Knox* et al., 1979]. Since poor socioeconomic societies tend to have children at a young age, exposure of their offspring is high. When newborns are exposed to virus in these sources, a sizable proportion will acquire infection [*Hayes* et al., 1972; *Leinikki* et al., 1978; *Stagno* et al., 1980]. Additional numbers may become infected by transmission

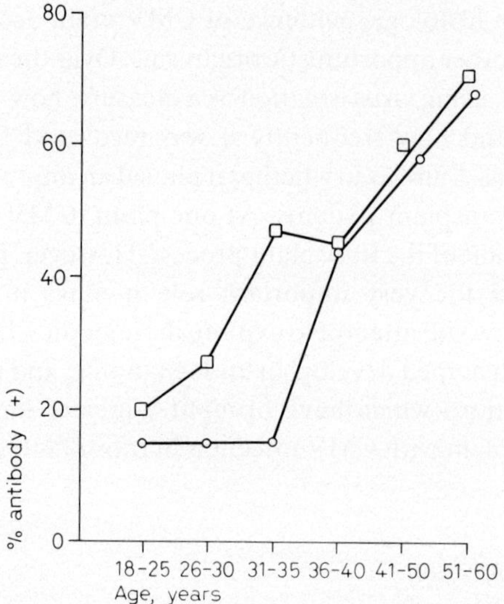

Fig. 1. Frequency of antibody to CMV in white females compared to white males. Percentage of antibody-positive individuals based on age and sex. □ = Females; ○ = males. Approximately 100 individuals of each sex were tested in each age group.

among children in the family and/or among playmates, although it is uncertain how frequently this occurs [*Hanshaw* et al., 1965]. By the time the children in this population reach teenage years, the majority have been infected. Those who have not will probably acquire infection related to sexual activity. Thus, by their mid-twenties, nearly 90% of the women and 80% of the men in this lower socioeconomic group are antibody-positive and have latent or active CMV infection [*Betts,* unpublished data].

In poor white populations, antibody studies reveal that the frequency of positivity is lower than in poor black populations but is higher than the frequency in middle class white populations [*Betts,* unpublished data]. Presumably the same factors contribute to infection among the young in the lower socioeconomic white populations as in the black population. The greatest difference among groups of young people exists between the black and the middle class white populations. In the latter group, frequency of antibody is much lower: 15% in men and 20% in women. Thus, there are definitely different frequencies of antibody among 20-year-old black women, 20-year-old lower class white women, and 20-year-old middle class women: 70, 40 and 20%, respectively. Men in these groups have a lower frequency of antibody than women, although studies involving males are much less frequent (fig. 1).

In spite of the fact that members of the middle class group have a low frequency of antibody at age 20, by age 60 the antibody frequency is high (70%). A variety of data suggests that the degree of sexual activity is a very important contributor to the steady increase in frequency with which infection is acquired during those years [*Jordan* et al., 1973; *Davis* et al., 1975]. Women are more likely to acquire infection between the ages of 20 and 35 years, while men are more likely to acquire infection after the age of 35 years (fig. 1). By the age of 50 or 60 years, the frequency of antibody positivity is similar in men and women. An exception to that rule occurs in homosexual males where antibody frequency among men in their 20s is very high. Interestingly, viruria in the homosexual male population is fairly common in the early 20s but decreases by the early 30s [*Drew* et al., 1980]. This pattern is very reminiscent of cervical virus shedding in the young antibody-positive female population.

Other factors may influence the rate of acquisition of antibody in addition to sex, race, and socioeconomic status. Among populations where teenagers have a low frequency of antibody, blood group appears to influence the rate at which young adults acquire antibody. Blood group A females are much more likely to have acquired antibody by age 35 than blood group O females. Males of either blood group, as suggested by the information mentioned above, are less likely to be antibody-positive [*Andrus* et al., 1981].

All in all, the information regarding the epidemiology of CMV infection indicates that different populations undergoing allotransplantation have a strikingly different frequency of latent infection with CMV at the time of transplantation. It is therefore necesssary to use this knowledge of the epidemiology of CMV when interpreting the studies of CMV and organ allotransplantation.

III. Types of Transplants

There are three organ systems in which allotransplantation has been extensively conducted and in which there is knowledge of infection with CMV. These are renal, bone marrow and cardiac allotransplantation. Presumably, some of the same features in these groups would be true for liver, pancreas and lung allotransplant recipients but there are no data to corroborate this presumption. Further remarks in this review will be confined to the three major types of transplants listed above.

IV. Pathogenesis of CMV Infection

The study of potential renal allograft recipients while they are undergoing several months or years of hemodialysis, demonstrates that little or no infection is acquired due to transmission on the dialysis unit [*Betts* et al., 1979; *Tolkoff-Rubin* et al., 1978]. Furthermore, a number of carefully conducted studies have demonstrated that no virus is detectable in secretions of either potential recipients or donors prior to renal transplantation [*Betts* et al., 1975; *Fiala* et al., 1975; *Ho* et al., 1975]. However, approximately 1 month following transplantation, virus shedding may be detectable and either may persist for several months thereafter or may be intermittently detectable. If the virus is not detectable prior to transplant, then what is the source of virus?

A. Source of CMV in Allograft Infection

A number of studies suggest that there are three sites from which CMV infection may be derived after transplantation. First, virus latent in the recipient prior to transplantation is reactivated by the transplant process and/or immunosuppressive drugs [*Betts* et al., 1975; *Fiala* et al., 1975; *Ho* et al., 1975; *Craighead* et al., 1967]. Second, virus in the transplanted allograft (either parenchymal cells or circulating cells trapped in the allograft) is reactivated by the same factors [*Betts* et al., 1975; *Ho* et al., 1975; *Pass* et al., 1978]. Third, latent virus in white blood cells transfused into the recipient is reactivated. In settings where only frozen blood is used, as in the hemodialysis and renal transplant population, blood transfusion does not appear to be an important source of CMV infection [*Betts* et al., 1979; *Tolkoff-Rubin* et al., 1978]. However, in patients who receive substantial volumes of fresh blood, as bone marrow transplants and cardiac transplants, blood transfusion is probably a very important source of CMV [*Winston* et al., 1980]. In fact, it may account for most virus infection in bone marrow transplants and most primary infection in cardiac transplants. These three sites are the major source of virus that infects these patients. Only very rarely is virus acquired from other infected individuals [*Betts* et al., 1978; *Tolkoff-Rubin* et al., 1978].

The information to support these three sources of CMV is almost exclusively epidemiologic in nature. For example (table I), renal allograft recipients who are seronegative (no latent infection) prior to transplantation and who receive a kidney from a donor who is also seronegative, never shed cyto-

Table I. Percent of renal allograft recipients who develop virus shedding after transplant in relation to pretransplant antibody status of donor and recipient

CMV antibody status of donor	CMV antibody status of recipient	
	+	–
+	≈ 90%	≈ 75%
–	≈ 90%	0

megalovirus after transplantation [*Betts* et al., 1975]. This is in contrast to those individuals who are seronegative and receive a kidney from a seropositive donor. Usually, but not always, they shed virus after transplant [*Betts* et al., 1975; *Ho* et al., 1975; *Pass* et al., 1978]. Patients who are seropositive and receive a kidney from a seronegative or a seropositive donor almost always shed virus, suggesting that a least most of the virus infection occurring in these individuals is due to reactivation of the recipient's own virus. The actual source of virus in the setting of seropositive donor-seropositive recipient is, of course, unclear, and molecular data are not yet available to determine whether these individuals may shed viruses of two different molecular types [*Huang* et al., 1980]. There are, however, data to suggest that pretransplant seropositive patients receiving kidney from seropositive donors are more likely to form IgM anti-CMV antibody post-transplant than if they received a kidney from a seronegative donor [*Betts and George,* 1980; *Betts and Schmidt,* 1981].

The data indicating a CMV source among bone marrow allograft recipients are not as clearly defined. Certainly, a donor allograft could contribute virus if the donor had latent infection, but other sources appear to be equally important [*Meyers* et al., 1975; *Winston* et al., 1980]. In one study where recipients received prophylactic white cell transfusions, CMV infection rate was strikingly higher than in patients who did not receive prophylactic white blood cells [*Winston* et al., 1980]. Recognizing that other blood transfusions given in the previous several months prior to bone marrow transplantation could also have transmitted CMV infection, it is easy to see why the incidence of infection in bone marrow recipients is so very high. Blood as a source of virus is an important consideration because in bone marrow transplantation many of the subjects are young, white and middle class, a group which normally has a very low incidence of latent CMV infection.

B. Reactivation of Latent Virus

What leads to reactivation of virus from any of these three sources? The answer to this major question is not completely understood, but studies in non-transplanted patients have been helpful. *Dowling* and co-workers have conducted studies in patients with rheumatoid arthritis [*Dowling* et al., 1976]. During the baseline period they infrequently shed CMV; however, 1–2 months after initiation of azothioprine or cyclophosphamide therapy, roughly one-third of the patients shed virus in their urine. Reactivation of latent virus by the cytotoxic therapy is the most likely explanation in these patients. In contrast, prednisone-treated patients did not appear to reactivate latent virus. Animal studies give support to these observations. Experimental mice with latent CMV, when given cyclophosphamide, reactivate that CMV. It is, however, also important to note that other studies in experimental animals have demonstrated that placing a skin allograft on an animal with latent murine CMV infection will lead to reactivation of that virus [*Wu* et al., 1975; *Jordan* et al., 1977; *Mayo* et al., 1977].

Patients who undergo transplantation generally receive azothioprine and prednisone with a foreign graft, factors that may interact to influence reactivation of latent CMV. Measurement of cell-mediated immune response to CMV shows drastic decrease in the period following both renal and cardiac transplant [*Linnemann* et al., 1978b; *Pollard* et al., 1978; *Rytel* et al., 1978]. Whatever the explanation for reactivation of virus may be, the process leads to reactivation not only of the patient's own latent virus but also of virus latent in transfused white blood cells and/or in the donated allograft. The fact that it is latent virus rather than active virus is substantiated by the fact that direct cultivation of blood units [*Armstrong* et al., 1976; *Betts* et al., 1979] or of explants of renal parenchymal cells has failed to reveal infectious virus [*Seidel* et al., 1978; *Naraqui* et al., 1978].

C. Types of Infection in Different Pairings

Considering the possible sources of virus and the epidemiologic factors mentioned above, one could predict what types of infection might occur, and the expected frequency of each type of infection, among transplant patients (table II, fig. 1).

In the renal transplant patient three types of pairings are possible. One type occurs between HLA-identical siblings and usually involves two rela-

Table II. Relationships of type of pairing and type of transplant procedure to the development of primary or reactivation CMV infection in transplant recipient

Renal allograft match	Expected antibody frequency		Frequency of infection	
	donor	recipient	primary	reactivation
HLA-identical match	low	low	very low	low
Haploidentical (parent-child)	high	low	high	low
Cadaveric	low	high	low	high
Bone marrow				
HLA-identical	low	high[1]	of donor marrow	high
Cardiac				
Cadaveric	low	high	low	high

[1] Due to blood products.

tively young subjects who are close to one another in age. In the HLA-identical pairings, siblings not only share HLA antigens but often have the same status with respect to latent CMV infection [*Betts* et al., 1977]. Remembering that the frequency of latent infection among white middle class individuals is low until their early 30s, HLA-identical donor and recipient pairs are often both CMV-negative. Similarly, when one is positive the other is usually also positive. Another type which involves live, related matches is transplantation betweeen haploidentical individuals. This usually involves parent-child pairings although child-to-parent is sometimes carried out. In the haploidentical pairing, because of the parents' older age (40–50 years), they have often developed latent infection whilc the child, generally 25 years or younger, is usually free of latent infection. This provides the setting for primary infection which is quite common in the haploidentical matches [*Betts* et al., 1977]. The third type is nonrelated donor-recipient pairings, almost always a cadaver donor. The recipient of a cadaver kidney is often too old to have an eligible parent to donate. This older age also means that the recipient has usually acquired latent infection as would an individual of equal age without renal disease. By contrast, most of the cadaver donors are quite young and very often male (young accident victims), and therefore usually free of latent infection. Primary infection in recipients of cadaver kidneys, then, is generally uncommon. Because of epidemiologic factors, these three types of pairings are distinctly different with respect to the type of infection that can be anticipated post-transplant. HLA-identical siblings have a high frequency of no infection, haploidenticals a high frequency of primary infection, and recipients of cadaver kidneys a high frequency of reactivation of their own latent virus. The exception would be when the pairings listed above are between black donor and recipient, both of whom are usually seropositive, for the reasons listed under Epidemiology. This situation, no matter what the pairing, would most likely result in reactivation of latent virus.

In bone marrow recipients the pairings should be similar to the HLA-identical population in the renal transplant program. Most marrow recipients are young, and the frequency of latent infection in donor and recipient should be low. However, for reasons already mentioned (multiple blood transfusions), the frequency of infection developing after the diagnosis of leukemia is established is quite high. By contrast, the donor who is often young and not exposed to blood transfusions has not previously experienced CMV infection. Thus, the donor is seldom the source of primary infection. Even though the donor may not be a source of virus, the graft itself may become infected with the recipient's virus. This unique biologic phenomenon may help explain some of the striking findings which develop in the bone marrow graft recipient. This point will be discussed later.

The recipient of the cardiac allograft is very similar to the cadaver recipient in the renal transplant setting with regard to the outcome of CMV in relation to the type of pairing. However, transfused blood may be an additional source for primary infection in the small number of recipients who do not have naturally acquired latent CMV infection at the time of transplantation.

V. Clinical Manifestations of Infection with Cytomegalovirus Following Transplantation

A. Febrile Illness Due to CMV

Features of the CMV syndrome in renal transplant patients are variable, including fever, change in white blood count (usually leukopenia but occasionally leukocytosis), liver enzyme increases (particularly SGOT), and pneumonitis [*Armstrong* et al., 1971; *Betts* et al., 1977; *Fiala* et al., 1975; *Fine* et al., 1972; *Ho* et al., 1975; *Luby* et al., 1974; *Pass* et al., 1978, 1980; *Rubin* et al., 1977; *Simmons* et al., 1974, 1977; *Spencer*, 1974; *Suwansirikul* et al., 1977; *Peterson* et al., 1980]. The x-ray in patients with pneumonitis proven by biopsy to be CMV may reveal a diffuse reticular pattern, bronchopneumonia or rarely a lobar pattern. Effusion has been noted as well [*Betts* et al., 1977; *Suwansirikul* et al., 1977]. The syndrome itself may not include all the elements (table III), and the range of some of the abnormalities (for example, white blood count) is very broad. Liver function abnormalities are very mild but can persist, and virus can be isolated from the liver. Almost by definition, fever is part of the syndrome, but in view of the fact that many more patients shed CMV than have fever, it may be that some of the abnormalities that are seen after transplant are due to CMV. In addition to the symptoms listed in table III, a smaller fraction of patients will also have arthralgia and/or arthritis [*Fiala* et al., 1975]. Other complaints are much less common. In some, infection has been shown to result in death [*Rifkind* et al., 1964; *Hedley-Whyte and Craighead,* 1965; *Simmons* et al., 1977; *Light and Burke,* 1979]. Although most of the symptoms due to CMV occur in the first 90 days after transplant, some patients will develop symptoms due to infection 1 or 2 years later [*Linnemann* et al., 1978a; *Betts,* unpublished data].

A gradual understanding of factors that predict patients who will develop symptoms following CMV infection has emerged. As noted above, virus infection may develop in a subject who is seronegative prior to undergoing allograft transplantation. In those individuals who have primary infection, the above syndrome is quite common and occasionally is quite severe [*Fine* et al., 1972; *Betts* et al., 1977; *Suwansirikul* et al., 1977; *Pass* et al., 1978]. If all other factors are equal, the frequency of severe disease is greater in those developing primary infection compared to those who have had antibody prior to allograft transplantation.

The majority of cases involving primary infection will occur in haploidentical matches for the epidemiologic reasons outlined previously. Therefore, primary infection most often occurs in younger individuals who are receiving a kidney that has a good chance of success. However, haploidentical matches are less frequent than allograft placements between cadaver donor and unrelated recipients. Among recipients of cadaveric kidneys the frequency of primary infection is very low. On a chance basis, 60% of cadaver recipients, because of their older age, are seropositive prior to transplant. Their donors are, because of their younger age, usually seronegative (80%). Therefore, when these pairings are carried out without regard to CMV status, one could anticipate that only 8 out of 100 cadaveric recipients would be at risk to develop primary infection [*Betts* et al., 1977; *May* et al., 1978; *Andrus* et al., 1981]. Almost all virus shedding that occurs in cadaveric recipients is due to reactivation of virus which was latent in the recipient at the time of transplantation. For cadaveric recipients a more important factor

Table III. Clinical parameters occurring in transplant-associated CMV illness

	Frequency, %	Severity
Fever	100[1]	100–104 °F, 1–4 weeks
Lymphadenopathy	0	
Pharyngitis	0	
Hepatomegaly	10–30	mild
Splenomegaly	5–40	2–3 finger breadths
Skin rash	5–10	associated with carbenicillin
Liver function abnormality	40	low grade
Leukopenia	40	<2,000
Lymphocytosis	5	>35,000
Pneumonia	10–15	hypoxia, focal or diffuse

[1] Required to be included.

which predicts development of illness associated with CMV infection is the type of immunosuppression administered. For example, symptoms associated with CMV infection are much more likely to occur in individuals receiving antithymocyte globulin (ATG) than in those who do not [*Cheeseman* et al., 1979; *Pass* et al., 1980]. This may be explained by the fact that ATG leads to a much greater T-cell depression, greater frequency of viremia and higher titers of virus than other types of immunosuppressive agents [*Cheeseman* et al., 1979; *Pass* et al., 1980; *Thomas* et al., 1976]. The bone marrow transplant recipient who undergoes total body irradiation and/or very high doses of cytotoxic agents prior to the receipt of a bone marrow allotransplant also is maximally immunosuppressed. If the donor allograft develops 'primary infection' from the recipient virus which is reactivated by the preparative chemotherapy, then alteration of the immune function of the bone marrow segment of the immune system might occur. This could alter the antigenic match between donor and recipient in a manner described by *Zinkernagel and Oldstone* [1976] in experimental animals. The HLA antigen which has incorporated foreign viral antigen might serve as a focus of attack by cytotoxic T cells. These speculations are currently unsubstantiated by data, but the high frequency of graft-versus-host disease and the association with CMV infection is intriguing [*Meyers* et al., 1975; *Leinikki* et al., 1978; *Winston* et al., 1980]. When CMV viremia is demonstrable in bone marrow recipients who have graft-versus-host disease and interstitial pneumonitis, death is the usual outcome.

B. Bacterial and Fungal Superinfection

In the renal transplant recipient with primary infection, recovery is the rule. This is true because it so often occurs in haploidentical matches where graft survival is excellent and immunosuppression often does not include ATG. However, in those individuals who develop leukopenia associated with CMV infection, subsequent bacterial and/or fungal infection may develop [*Rubin* et al., 1977; *Chatterjee* et al., 1978]. One interesting organism that has produced infection in this setting is *Listeria monocytogenes.*

The occurrence of bacterial superinfection associated with CMV may be related to local tissue effects produced by virus, allowing bacterial invasion, or may be related to interference in the host immune response [*Kelsey* et al., 1977; *Linnemann* et al., 1978b; *Rytel* et al., 1978; *Hamilton* et al., 1977]. For example, the pulmonary macrophage is one of the few non-fibroblastic cells that support the growth of CMV in vitro. Since the pulmonary macrophages may provide an important line of host defense against bacterial infection of the lung and if replication of CMV in these cells occurs in vivo, temporary incapacity of these important cellular elements due to infection by CMV might explain the observations which have been made. Effect on pulmonary macrophages, combined with the peripheral leukopenia and the mechanical problems of congestive heart failure, may explain the association of bacterial infection in the lungs with primary CMV infection in cardiac transplant recipients [*Rand* et al., 1978].

C. Allograft Failure

Perhaps the greatest area of disagreement regarding the various problems associated with CMV is whether or not the virus can contribute to allograft failure in renal transplant patients. This is an extremely complex area. Recognizable CMV infection as outlined above is likely to occur in two settings: following primary infection and in individuals receiving ATG. Primary infection predominates in haploidentical parent-child pairings, a pairing which in general has a favorable outcome. Since ATG is primarily used in recipients of cadaver kidneys, symptoms associated with ATG in reactivation infection are much more common in a situation in which the outcome for other reasons is relatively poor. Furthermore, if ATG has a beneficial effect on allograft survial [*Cosimi* et al., 1976; *Michel* et al., 1975] but a detrimental effect on virus infection [*Cheeseman* et al., 1979], these two factors might neutralize one another with respect to overall outcome.

If this were the only problem which made analysis difficult, an answer might still be achievable. However, other factors apparently play a role in transplant outcome. For example, the HLA match may exert an influence on allograft success, as does administration of more than five blood transfusions prior to transplantation [*Briggs* et al., 1978; *Fuller* et al., 1978; *Opelz and Terasaki,* 1976, 1978; *van Es and Balner,* 1979; *Vincenti* et al., 1978]. In some studies, nonreactivity in the mixed lymphocyte culture is associated with increased allograft success [*Vincenti* et al., 1978], and in other studies ABO blood group O patients have a better outcome than ABO-A or -B patients [*Bore* et al., 1979; *Burleson* et al., 1978]. Furthermore, the method of preserving the cadaveric allograft before transplant also has an influence.

In order to judge the effect of CMV on outcome of the allograft it is important to take into account these other factors as well [*Andrus* et al., 1981].

One other reason that there have been differences between studies and thus disagreement in results has been that there is not always uniformity in what is considered to be CMV infection. For example, some studies have analyzed the effect of CMV infection on allograft outcome by comparing patients with 'overt' CMV infection versus all other patients [*Peterson* et al., 1980]. However, the symptoms of overt infection may be caused by an agent other than CMV. Secondly, there are patients with CMV infection who do not fall in the overt group yet they are actively shedding CMV and may even have viremia. They fall into the 'other' group or non-CMV group. Other studies require that virus shedding be present for the patient to be included, but virus shedding can be intermittent [*Betts* et al., 1975]. Still other studies have utilized presence or absence of antibody as a measure of infection. Clearly, the individuals who would be included in the non-CMV group would be different in each of these three types of studies.

In a study where antibody was taken as a measure of CMV infection among cadaveric recipients and where little or no ATG was administered, presence of CMV antibody prior to transplant resulted in a more unfavorable outcome than its absence [*May* et al., 1978]. If these groups were subdivided according to whether or not multiple units of frozen blood (vide supra) were administered prior to transplantation, there appeared to be a relationship between blood transfusion and CMV status [*Andrus* et al., 1979]. The individuals without latent infection who received multiple transfusions had the best outcome, those with latent infection who received multiple transfusions had an intermediate outcome, and those who did not receive transfusions had the poorest outcome, independent of infection status. There was no obvious effect of race, HLA match, sex or age on these figures.

It is intriguing to speculate that the studies which have shown an advantage of blood group O over group A could be also showing that CMV has an influence on the outcome since blood group O subjects may have a lower incidence of latent CMV infection than other blood types [*Andrus* et al., 1981]. Studies that have failed to show the advantage of blood group O come from the southeastern area of the United States where there is a very high incidence of latent CMV infection in all blood groups.

A mechanism by which CMV might contribute to allograft failure in transplant recipients has recently been suggested [*Richardson* et al., 1981]. Patients developing CMV viremia after transplantation developed renal dysfunction simultaneous to the development of that viremia. Histologically, a diffuse glomerulopathy was noted in the patients. Renal dysfunction improved with the decrease or cessation of immunosuppressive therapy. Other patients with renal dysfunction had typical changes of rejection and did not have viremia. The implication is that viremia leads to immune complex deposition and to renal dysfunction in some patients. This interesting observation bears further study.

It is clear that in order to gain an understanding of the role of CMV in allograft failure, all factors must be considered. Further studies need to define the study group carefully. Patients with CMV infection should not be excluded from the infected group simply because the infection is not 'overt'.

VI. Special Problems

There are special problems with CMV in allograft recipients which are of importance, although not unique to allograft recipients. For instance, retinitis due to CMV is more common than in other compromised hosts [*Pollard* et al., 1980]. Although the extent of disease is variable, it can be very severe. The diagnosis is made on clinical grounds when the lesions occur. CMV viruria and an increase in antibody titer also support the diagnosis but are not unique to this entity.

Interstitial pneumonitis in bone marrow transplant recipients is another problem in which the relationship of CMV is not totally established. Proof that CMV causes pneumonitis is not yet complete but the suspicion is very strong [*Neiman* et al., 1973; *Meyers* et al., 1975]. The temporal sequence is quite repetitive; viremia in the 3-month period following transplant is followed within 3–4 weeks by pneumonitis. When this occurs, the death rate is very high. The course is usually quite fulminant and may be associated with elements of graft-versus-host disease such as hyperbilirubinemia [*Meyers* et al., 1975; *Winston* et al., 1980].

VII. Prevention and Treatment

A. Donor Selection

Although no formal studies have been conducted, there are data to suggest that if seronegative potential renal allograft patients receive only previously frozen red blood transfusions and a graft from a seronegative donor, primary infection can be avoided [*Betts* et al., 1977; *Tolkoff-Rubin* et al., 1978; *Betts* et al., 1979]. This solution is not entirely acceptable. As mentioned in the section on types of infection in different pairings, the most frequent setting in which a positive donor is matched to a negative recipient is in haploidentical parent-child matches. If these transplants were not carried out, then a sizable number of kidneys would be removed from the donor pool. Because of limited kidneys for donation and the relative success of haploidentical matches, eliminating such matches is not feasible. Because of the epidemiologic circumstances referred to previously, the possibility of choosing an organ from a donor who has no latent CMV infection is quite feasible. Therefore, this type of donor would be the most logical choice for many dialysis and cardiac patients awaiting transplant. Leukemia patients who may eventually undergo bone marrow transplant, who must receive whole blood or other non-frozen blood elements, would most definitely benefit from CMV-seronegative donors, thereby reducing problems related to CMV infection [*Winston* et al., 1980].

B. Vaccine

It has been proposed that vaccine might provide an opportunity to avoid infection and/or illness from virus in the transplanted allograft. The only study currently available [*Glazer* et al., 1979] has shown no protective effect against infection and no obvious reduction in illness among seronegative vaccinated recipients who receive a kidney from a seropositive donor. This lack of benefit may be explained by the fact that the immunosuppression regimens drastically reduce not only vaccine-induced but also natural immunity to CMV just at the time when it is needed most [*Linnemann* et al.,

1978b; *Pollard* et al., 1978]. Currently, a larger study is underway to determine any benefit of vaccine against illness. The results of these studies are awaited.

C. Passive Immunity

Because the aim in the population in which the infection develops is to keep an allograft in place, passive immunity using cells that would recognize viral antigen does not seem feasible.

Using serum and/or plasma theoretically could be helpful, and observations in neonates support that contention [*Yeager* et al., 1981]. Since the antibody that would probably be useful in this setting is one that demonstrates a capacity to neutralize, studies should be done in which this parameter is measured. The only studies currently available have been conducted using serum in which only complement-fixing titer has been measured.

D. Antiviral Chemotherapy

Although there have been very definite advances in the use of antiviral therapy against other herpesviruses, progress against CMV has been much slower. While there is some hint that adenine arabinoside [*Pollard* et al., 1980] or interferon [*Cheeseman* et al., 1979; *Meyers* et al., 1980] or both, may be helpful, the benefits are marginal or the toxicity is unacceptable [*Marker* et al., 1980]. Acyclovir, the newer compound being studied, has no obvious therapeutic effect against CMV. However, use of this antiviral agent prophylactically, i.e., in the bone marrow transplant recipient, is now under study.

VIII. Summary

Cytomegalovirus infection associated with allograft transplantation is one of the better studied virus infections in this setting. The type of infection which develops is largely explained by epidemiologic factors and by the fact that CMV can be transmitted by blood products and by the allograft. There is a well-recognized symptom complex associated with CMV infection which may occasionally be mistaken for bacterial infection or allograft rejection.

Important features associated with CMV infection are secondary bacterial or fungal infection due to leukopenia or to alteration in host defense mechanisms, or both; graft-versus-host disease in bone marrow allograft recipients; and acute rejection or allograft loss in renal allograft recipients.

Other than seronegative donor selection for seronegative potential allograft recipients, there is no means available to prevent or treat CMV infection at this time. Further studies with antiviral agents, interferon, passive immunity or combinations thereof seem justified in view of the extent of the problems associated with CMV infection.

References

Andrus, C.H.; Betts, R.F.; May, A.G.; Freeman, R.B.: Cytomegalovirus infection blocks the beneficial effect of pretransplant blood transfusion on renal allograft survival. Transplantation *28:* 451–456 (1979).

Andrus, C.H.; Betts, R.F.; May, A.G.; Freeman, R.B.: Better allograft survival in erythrocyte Type-O recipients correlates with resistance to cytomegalovirus infections. Transplant. Proc. *XIII:* 120–124 (1981).

Armstrong, D.; Balakrishnan, S.L.; Steger, L.; Yu, B.; Stenzel, K.H.: Cytomegalovirus infections with viremia following renal transplantation. Archs intern. Med. *127:* 111–116 (1971).

Armstrong, J.A.; Tarr, G.C.; Youngblood, L.A.; Dowling, J.N.; Saslow, A.R.; Lucas, J.P.; Ho, M.: Cytomegalovirus infection in children undergoing open-heart surgery. Yale J. Biol. Med. *49:* 83–91 (1976).

Betts, R.F.; Cestero, R.V.M.; Freeman, R.B.; Douglas, R.G., Jr.: Epidemiology of cytomegalovirus infection in end stage renal disease. J. med. Virol. *4:* 89–96 (1979).

Betts, R.F.; Freeman, R.B.; Douglas, R.G., Jr.; Talley, T.E.: Clinical manifestations of renal allograft derived primary cytomegalovirus infection. Am. J. Dis. Child. *131:* 759–763 (1977).

Betts, R.F.; Freeman, R.B.; Douglas, R.G., Jr.; Talley, T.E.; Rundel, B.: Transmission of cytomegalovirus infection with renal allograft. Kidney int. *8:* 387–394 (1975).

Betts, R.F.; Schmidt, S.D.: Evaluation of cytolytic antibody to cytomegalovirus infected cells in patients undergoing renal transplantation; in Transplantation and clinical immunology XII, p. 55 (Excerpta Medica, Amsterdam 1980).

Betts, R.F.; George, S.D.: Cytolytic IgM anticytomegalovirus antibody in primary cytomegalovirus infection in man. J. infect. Dis. *143:* 821–826 (1981).

Bore, P.J.; Sells, R.A.; Jameson, V.: Transfusion-induced renal allograft protection. Transplant. Proc. *11:* 148 (1979).

Briggs, J.D.; Canavan, J.S.F.; Dick, H.M.; Hamilton, D.N.H.; Kyle, K.F.; Macpherson, S.G.; Paton, A.M.; Titterington, D.M.: Influence of HLA matching and blood transfusion on renal allograft survival. Transplantation *25:* 80–85 (1978).

Burleson, R.L.; Stahl, P.J.; Henry, J.B.: Renal transplantation. N.Y. St. J. Med. *78:* 2019 (1978).

Cappel, R.; Hestermans, O.; Toussaint, C.; Vereerstraeten, P.; van Beers, D.; DeBraekeleer, J.; Schoutens, E.: Cytomegalovirus infection and graft survival in renal graft recipients. Archs Virol. *56:* 149–156 (1978).

Chatterjee, S.N.; Fiala, M.; Weiner, J.; Stewart, J.A., Stacey, B.; Warner, N.: Primary cytomegalovirus and opportunistic infections. J. Am. med. Ass. *240:* 2446–2449 (1978).

Cheeseman, S.H.; Rubin, R.H.; Stewart, J.A.; Tolkoff-Rubin, N.E.; Cosimi, A.B.; Cantell, K.; Gilbert, J.; Winkle, S.; Herrin, J.T.; Black, P.H.; Russell, P.S.; Hirsch, M.S.: Controlled clinical trial of prophylactic human-leukocyte interferon in renal transplantation. New Engl. J. Med. *300:* 1345–1349 (1979).

Cosimi, A.B.; Wortis, H.H.; Delmonico, F.L.; Russell, P.S.: Randomized clinical trial of antithymocyte globulin in cadaver renal allograft recipients: importance of T cell monitoring. Surgery, St Louis *80:* 155–163 (1976).

Craighead, J.E.; Hanshaw, J.B.; Carpenter, C.B.: Cytomegalovirus infection after renal allotransplantation. J. Am. med. Ass. *201:* 99–102 (1967).

Davis, L.E.; Stewart, J.A.; Garvin, S.: Cytomegalovirus infection: a seroepidemiologic comparison of nuns and women from a venereal disease clinic. Am. J. Epidem. *102:* 327–330 (1975).

Dowling, J.N.; Saslow, A.R.; Armstrong, J.A.; Ho, M.: Cytomegalovirus infection in patients receiving immunosuppressive therapy for rheumatologic disorders. J. infect. Dis. *133:* 399–408 (1976).

Drew, W.L.; Mintz, L.; Hoo, R.; Finley, T.N.: Growth of herpes simplex and cytomegalovirus in cultured human alveolar macrophages. Am. Rev. resp. Dis. *119:* 287–291 (1979).

Drew, W.L.; Mintz, L.; Sands, M.; Minter, R.C.; Ketterer, B.: Cytomegalovirus infection in homosexual men. Abstr. Am. Soc. Microbiol., Interscience Conference on Antimicrobial Agents and Chemotherapy, No. 386 (Sept. 22–24, 1980).

Fiala, M.; Payne, J.E.; Berne, T.V.; Moore, T.C.; Henle, W.; Montgomerie, J.Z.; Chatterjee, S.N.; Guze, L.B.: Epidemiology of cytomegalovirus infection after transplantation and immunosuppression. J. infect. Dis. *132:* 421–433 (1975).

Fine, R.N.; Grushkin, C.M.; Malekzadeh, M.; Wright, H.T., Jr.: Cytomegalovirus syndrome following renal transplantation. Archs Surg., Chicago *105:* 564–570 (1972).

Fuller, T.C.; Delmonico, F.L.; Cosimi, A.B.; Huggins, C.E.; King, M.; Russell, P.S.: Impact of blood transfusion on renal transplantation. Ann. Surg. *187:* 211–218 (1978).

Glazer, J.P.; Friedman, H.M.; Grossman, R.A.; Starr, S.E.; Barker, C.F.; Perloff, L.J.; Huang, E.-S.; Plotkin, S.A.: Live cytomegalovirus vaccination of renal transplant candidates. Ann. intern. Med. *91:* 676–683 (1979).

Hamilton, J.R.; Overall, J.C., Jr.; Glasgow, L.A.: Synergistic infection with murine cytomegalovirus and *Candida albicans* in mice. J. infect. Dis. *135:* 918–924 (1977).

Hanshaw, J.B.; Betts, R.F.; Simons, G.; Boynton, R.C.: Acquired cytomegalovirus infection. New Engl. Med. *272:* 602–609 (1965).

Hayes, K.; Danks, D.M.; Gibas, H.; Jack, I.: Cytomegalovirus in human milk. New Engl. J. Med. *287:* 177–178 (1972).

Hedley-Whyte, E.T.; Craighead, J.E.: Generalized cytomegalovirus inclusion disease after renal homotransplantation. New Engl. J. Med. *272:* 473–475 (1965).

Ho, M.; Suwansirikul, S.; Dowling, J.N.; Youngblood, L.A.; Armstrong, J.A.: The transplanted kidney as a source of cytomegalovirus infection. New Engl. J. Med. *293:* 1109–1112 (1975).

Huang, E.-S.; Alford, C.A.; Reynolds, D.W.; Stagno, S.; Pass, R.F.: Molecular epidemiology of cytomegalovirus infections in women and their infants. New Engl. J. Med. *303:* 958–962 (1980).

Jordan, M.C.; Rousseau, W.E.; Noble, G.R.; Stewart, J.A.; Chin, T.D.Y.: Association of cervical cytomegaloviruses with venereal disease. New Engl. J. Med. *288:* 932–934 (1973).

Jordan, M.C.; Shanley, J.D.; Stevens, J.G.: Immunosuppression reactivates and disseminates latent murine cytomegalovirus. J. gen. Virol. *37:* 419–423 (1977).

Kelsey, D.K.; Olsen, G.A.; Overall, J.C., Jr.; Glasgow, L.A.: Alteration of host defense mechanisms by murine cytomegalovirus infection. Infect. Immunity *18:* 754–760 (1977).

Knox, G.E.; Pass, R.F.; Reynolds, D.W.; Stagno, S.; Alford, C.A.: Comparative prevalence of subclinical cytomegalovirus and herpes simplex virus infections in the genital and urinary tracts of low-income, urban women. J. infect. Dis. *140:* 419–422 (1979).

Leinikki, K.; Granstrom, M.-L.; Santavuori, P.; Pettay, O.: Epidemiology of cytomegalovirus infections during pregnancy and infancy. Scand. J. infect. Dis. *10:* 165–171 (1978).

Li, F.; Hanshaw, J.B.: Cytomegalovirus among migrant children. Am. J. Epidem. *86:* 137–141 (1967).

Light, J.A.; Burke, D.S.: Association of cytomegalovirus (CMV) infection with increased recipient mortality following transplantation. Transplant. Proc. *XI:* 79–82 (1979).

Linnemann, C.C.; Dunn, C.R.; First, M.R.; Alvira, M.; Schiff, G.M.: Late onset of fatal cytomegalovirus infection after renal transplantation. Archs intern. Med. *138:* 1247–1250 (1978a).

Linnemann, C.C., Jr.; Kauffman, C.A.; First, M.R.; Schiff, G.M.; Phair, J.P.: Cellular immune response to cytomegalovirus infection after renal transplantation. Infect. Immunity *22:* 176–180 (1978b).

Luby, J.P.; Barnett, W.; Hull, A.R.; Ware, A.J.; Shorey, J.W.; Peters, P.C.: Relationship between cytomegalovirus and hepatic function abnormalities in the period after renal transplant. J. infect. Dis. *129:* 511–518 (1974).

Luby, J.P.; Shasby, D.M.: A sex difference in the prevalence of antibodies to cytomegalovirus. J. Am. med. Ass. *222:* 1290–1291 (1972).

Marker, S.C.; Howard, R.J.; Groth, K.E.; Mastri, A.R.; Simmons, R.L.; Balfour, H.H.: A trial of vidarabine for cytomegalovirus infection in renal transplant patients. Archs intern. Med. *140:* 1441–1444 (1980).

May, A.G.; Betts, R.F.; Freeman, R.B.; Andrus, C.H.: An analysis of cytomegalovirus infection and HLA antigen matching on the outcome of renal transplantation. Ann. Surg. *187:* 110–117 (1978).

Mayo, D.R.; Armstrong, J.A.; Ho, M.: Reactivation of murine cytomegalovirus by cyclophosphamide. Nature, Lond. *267:* 721–723 (1977).

Meyers, J.D.; McGuffin, R.W.; Neiman, P.E.; Singer, J.W.; Thomas, E.D.: Toxicity and efficacy of human leukocyte interferon for treatment of cytomegalovirus pneumonia after marrow transplantation. J. infect. Dis. *141:* 555–562 (1980).

Meyers, J.D.; Spencer, H.C.; Watts, J.C.; Gregg, M.B.; Stewart, J.A.; Troupin, R.H.; Thomas, E.D.: Cytomegalovirus pneumonia after human marrow transplantation. Ann. intern. Med. *82:* 181–188 (1975).

Michel, R.P.; Guttman, R.D.; Knaack, J.; Klassen, J.; Beaudoin, J.-G.; Morehouse, D.D.: Antilymphocyte globulin in renal transplantation. Archs Surg., Chicago *110:* 90–93 (1975).

Naraqui, S.; Jackson, G.G.; Jonasson, O.; Rubenis, M.: Search for latent cytomegalovirus in renal allografts. Infect. Immunity *19:* 699–703 (1978).

Neiman, P.; Wasserman, P.B.; Wentworth, B.B.; Kao, G.G.; Lerner, K.G.; Storb, R.; Buckner, C.C.; Clift, R.A.; Fefer, A.; Fass, L.; Glucksberg, H.;Thomas, E.D.: Interstitial pneumonia and cytomegalovirus infection as complications of human marrow transplantation. Transplantation *15:* 478–485 (1973).

Opelz, G.; Terasaki, P.I.: Prolongation effect of blood transfusions on kidney graft survival. Transplantation *22:* 380–383 (1976).

Opelz, G.; Terasaki, P.I.: Improvement of kidney-graft survival with increased numbers of blood transfusions. Kidney Transpl. Blood Transf. *299:* 799–803 (1978).

Pass, R.F.; Long, W.K.; Whitley, R.J.; Soong, S.-J.; Diethelm, A.G.; Reynolds, D.W.; Alford, C.A., Jr.: Productive infection with cytomegalovirus and herpes simplex virus in renal transplant recipients: role of source of kidney. J. infect. Dis. *137:* 556–563 (1978).

Pass, R.F.; Whitley, R.J.; Diethelm, A.G.; Whelchel, J.D.; Reynolds, D.W.; Alford, C.A.: Cytomegalovirus infection in patients with renal transplants: potentiation by antithymocyte globulin and an incompatible graft. J. infect. Dis. *142:* 9–17 (1980).

Peterson, P.K.; Balfour, H.H.; Marker, S.C.; Fryd, D.S.; Howard, R.J.; Simmons, R.L.: Cytomegalovirus disease in renal allograft recipients: A prospective study of the clinical features, risk factors, and impact on renal transplantation. Medicine *59:* 283–300 (1980).

Pollard, R.B.; Egbert, P.R.; Gallagher, J.G.; Merigan, T.C.: Cytomegalovirus retinitis in immunosuppressed hosts. Ann. intern. Med. *93:* 655–670 (1980).

Pollard, R.B.; Rand, K.H.; Arvin, A.M.; Merigan, T.C.: Cell-mediated immunity to cytomegalovirus infection in normal subjects and cardiac transplant patients. J. infect. Dis. *137:* 541–549 (1978).

Rand, K.H.; Pollard, R.B.; Merigan, T.C.: Increased pulmonary superinfections in cardiac-transplant patients undergoing primary cytomegalovirus infection. New Engl. J. Med. *298:* 951–953 (1978).

Reynolds, D.W.; Stagno, S.; Hosty, T.S.; Tiller, M.; Alford, C.A., Jr.: Maternal cytomegalovirus excretion and perinatal infection. New Engl. J. Med. *289:* 1–5 (1973).

Richardson, W.P.; Colvin, R.B.; Cheeseman, S.H.; Tolkoff-Rubin, N.E.; Herrin, J.T.; Cosimi, A.B.; Collins, A.B.; Hirsh, M.S.; McCluskey, R.T.; Russell, P.S.; Rubin, R.H.: Glomerulopathy associated with cytomegalovirus viremia in renal allografts. New Engl. J. Med. *305:* 57–62 (1981).

Rifkind, D.; Starzl, T.E.; Marchino, T.L.; Waddell, W.R.; Rowlands, D.T., Jr.; Hill, R.B., Jr.: Transplantation pneumonia. J. Am. med. Ass. *189:* 808–812 (1964).

Rubin, R.H.; Cosimi, A.B.; Tolkoff-Rubin, N.E.; Russell, P.S.; Hirsch, M.S.: Infectious disease syndromes attributable to cytomegalovirus and their significance among renal transplant recipients. Transplantation *24:* 458–464 (1977).

Rytel, M.W.; Aguilar-Torres, F.G.; Balay, J.; Heim, L.R.: Assessment of the status of cell-mediated immunity in cytomegalovirus-infected renal allograft recipients. Cell. Immunol. *37:* 31–40 (1978).

Seidel, M.V.; Howard, R.J.; Balfour, H.H., Jr.: A virological study of human kidney explant cultures from renal allograft recipients. Transplantation *25:* 193–196 (1978).

Simmons, R.L.; Lobez, C.; Balfour, H., Jr.; Kalis, J.; Rattazzi, L.C.; Najarian, J.S.: Cytomegalovirus: clinical virological correlations in renal transplant recipients. Ann. Surg. *180:* 623–634 (1974).

Simmons, R.L.; Matas, A.J.; Rattazzi, L.C.; Balfour, H.H., Jr.; Howard, R.J.; Najarian, J.S.: Clinical characteristics of the lethal cytomegalovirus infection following renal transplantation. Surgery, St Louis *82:* 537–546 (1977).

Spencer, E.S.: Clinical aspects of cytomegalovirus infection in kidney-graft recipients. Scand. J. infect. Dis. *6:* 315–323 (1974).

Stagno, S.; Reynolds, D.W.; Pass, R.F.; Alford, C.A.: Breast milk and the risk of cytomegalovirus infection. New Engl. J. Med. *302:* 1073–1076 (1980).

Suwansirikul, S.; Rao, N.; Dowling, J.N.; Ho, M: Primary and secondary cytomegalovirus infection. Archs intern. Med. *137:* 1026–1029 (1977).

Thomas, F.; Lee, H.M.; Wolf, J.S.; Mendez-Picon, G.; Thomas, J.: Monitoring and modulation of immune reactivity in human transplant recipients. Surgery, St Louis *79:* 408–413 (1976).

Tolkoff-Rubin, N.E.; Rubin, R.H.; Keller, E.E.; Baker, G.P.; Stewart, J.A.; Hirsch, M.S.: Cytomegalovirus infection in dialysis patients and personnel. Ann. intern. Med. *89* (Part 1): 625–628 (1978).

van Es, A.A.; Balner, H.: Effect of pretransplant transfusion on kidney allograft survival. Transplant. Proc. *XI:* 127–137 (1979).

Vincenti, F.; Duca, R.M.; Amend, W.; Perkins, H.A.; Cochrum, K.C.; Feduska, N.J.; Salvatierra, O., Jr.: Immunological factors determining survival of cadaver-kidney transplants. New Engl. J. Med. *299:* 793–798 (1978).

Winston, D.J.; Ho, W.G.; Young, L.S.; Gale, R.P.: Prophylactic granulocyte transfusions during human bone marrow transplantation. Am. J. Med. *68:* 893–897 (1980).

Winston, D.J.; Meyer, D.V.; Gale, R.P.; Young, L.S.: Further experience with infections in bone marrow transplant recipients. Transplant. Proc. *X:* 247–254 (1978).

Wu, B.C.; Dowling, J.N.; Armstrong, J.A.; Ho, M.: Enhancement of mouse cytomegalovirus infection during host-versus-graft reaction. Science *190:* 56–58 (1975).

Yeager, A.S.; Grumet, F.C.; Hafleigh, E.B.; Arvin, A.M.; Bradley, J.S.; Prober, C.G.: Prevention of Tx-acquired cytomegalovirus infections in newborn infants. J. Pediat. *98:* 281–287 (1981).

Zinkernagel, R.M.; Oldstone, M.B.A.: Cells that express viral antigens but lack H-2 determinants are not lysed by immune thymus-derived lymphocytes but are lysed by other antiviral immune attack mechanisms. Immunology *73:* 3666–3670 (1976).

Dr. Robert F. Betts, Infectious Diseases Unit, Department of Medicine, University of Rochester School of Medicine, 601 Elmwood Avenue, Rochester, NY 14642 (USA)

Prog. med Virol., vol. 28, pp. 65–95 (Karger, Basel 1982)

Measles and SSPE Viruses: Similarities and Differences

Steven L. Wechsler [a], *H. Cody Meissner* [b]

[a]Department of Molecular Virology, Christ Hospital Institute of Medical Research, Cincinnati, Ohio; [b]Department of Microbiology and Molecular Genetics, Harvard Medical School, Boston, Mass., USA

Contents

I. Introduction

Measles, one of the classic diseases of childhood, is an acute, self-limited illness, caused by measles virus. Measles virus was first isolated in 1954 by *Enders and Peebles* [1]. An effective vaccine against measles was developed in 1960 by *Enders* et al. [2] and introduced for routine use in 1963. The undertaking of a successful measles vaccination program has greatly altered the incidence of clinical measles infections, reducing the incidence to about 10% of the 500,000 cases per year seen in the prevaccine era. However, measles remains an important area of investigation. This is due to a rare complication of measles virus infection, subacute sclerosing panencephalitis (SSPE) and to the tendency of measles virus to establish a persistent infection in a number of well-established cell lines. SSPE is of particular interest because it is one of the best studied examples in man of a persistent infection caused by a virus that normally produces an acute disease [see references 3–6 for reviews].

The earliest evidence that the agent of SSPE might be related to measles virus was a report in 1965 which described paramyxovirus nucleocapsids in electron micrographs of brain tissue taken from patients dying with SSPE [7]. Subsequent reports confirmed this finding [8–15]. Within 2 years after the initial report, elevated measles antibody titers in the serum and in the CSF of patients with SSPE were described and measles antigens were detected in SSPE brain tissue by immunofluorescent staining [9, 16]. Shortly afterward, infectious measles virus was recovered from the brain tissue of several SSPE patients, by use of co-cultivation techniques [17-23].

The brain tissue from patients dying with SSPE shows inflammatory lesions in both the white matter and the gray matter with occasional involvement of the cerebellum and the spinal cord [24–26]. While there may be varying degrees of demyelination, this extensive involvement of the brain defines a panencephalitis rather than the more restricted involvement seen in multiple sclerosis (a true demyelinating disease). The most characteristic finding, in SSPE brain cells, are the eosinophilic intranuclear inclusions [27] that have been demonstrated by electron microscopy to contain paramyxovirus-like particles [8–15].

Most data suggest that a measles infection in a child who later develops SSPE does not differ from the normal spectrum of measles in all children. The epidemiologic characteristics which have been associated with SSPE include the following: a 2- to 3-fold higher incidence in males over females; an increased incidence in children from rural areas, especially farms; and in the United States, far greater frequency of illness in the southeast [28, 29]. Perhaps the most striking association, however, is the observation that SSPE patients often have a history of contracting measles at a young age. The mean age of measles in SSPE patients is close to 2 years, while the mean age for all children in the United States is 4 years and for children from rural areas, 5 years. The latent period between measles and the onset of the earliest neurologic changes of SSPE has a mean of 6–7 years [28–31].

In the 12 years since measles virus was shown to be closely involved with SSPE, research efforts have been directed at defining those properties which permit a lytic, rapidly replicating virus to persist in a latent state. The intention of this article is to discuss and compare some of the physical and biological characteristics of measles and SSPE viruses. It will become clear that the hope of finding specific differences between the two viruses, which might explain the mechanism of persistence, has thus far not been realized.

It is important to stress that comparative studies of SSPE and measles viruses are difficult to interpret. The measles-like virus found in SSPE brain cells is non-productive and non-lytic. In the brain, no detectable extracellular viral particles are produced. However, the virus which is occasionally recovered from these brain cells by co-cultivation (designated SSPE virus) is usually productive and lytic in tissue culture. The co-cultivation procedures employed to recover SSPE virus from brain cells exert selective pressure for the recovery of a lytic virus that grows well in cells permissive for measles virus. In addition, the presence of virus is usually scored by looking for measles-virus-like cytopathic effects (CPE). Thus, it is not surprising that these infectious SSPE virus particles often have growth characteristics similar to wild-type measles virus.

Attempts to recover a virus from the brain tissue of patients dying with SSPE have been only partly successful [32]. The only instances in which SSPE viruses have been recovered from brain cells have followed extensive co-cultivation of brain cells with cells permissive for measles virus. Even then, in the majority of attempts (approximately 80% of the time), it has not been possible to recover virus [32]. Thus, when SSPE virus is rescued from brain cells of an individual with SSPE, it may represent either a minor population of the virus present or a newly arising mutant that occurs during the co-cultivation procedure. In either case, the lytic virus recovered and designated SSPE virus is different from the non-lytic virus originally present. Thus, care must be exercised when trying to draw conclusions about the pathogenesis of SSPE from comparative studies of SSPE viruses and measles viruses.

In addition to lytic SSPE viruses, other SSPE viruses are occasionally recovered as a persistent infection of the permissive cell line used for the co-cultivation procedure [22, 33–36]. This non-lytic SSPE virus remains strictly cell-associated (cell-associated SSPE), and little or no infectious virus is produced. It is possible that this virus more closely reflects the characteristics of the virus in the brain cells of SSPE patients. Indeed, when injected into animals, some of these cell-associated SSPE viruses are capable of producing a subacute encephalitis that is similar to SSPE in man [37–40]. However, like the lytic SSPE viruses discussed above, cell-associated SSPE viruses may have undergone some degree of selection and may differ from the virus in the brain cells of individuals with SSPE. Analysis of their characteristics is therefore difficult to interpret in terms of the original SSPE infection.

Thus, SSPE viruses have been isolated both as productive, lytic viruses and as non-productive, cell-associated viruses. Biologically, there does not seem to be any clear distinction between lytic SSPE viruses and lytic measles viruses. In addition, there does not appear to be any consistent difference between cell-associated SSPE viruses and cell-associated persistent measles virus infections. The important differences appear to be between those viruses that are cell-associated and those viruses that are lytic, regardless of whether they are SSPE viruses or measles viruses. Investigation into the differences between lytic and cell-associated viruses may eventually lead to a better understanding of measles virus persistence and the pathogenesis of SSPE.

II. The Diseases

Measles is a readily recognized illness that begins with conjunctivitis, cough and a characteristic rash. The incubation period is generally 10–12 days, and the illness usually resolves in less than 2 weeks. While measles is a self-limited illness with a very low mortality rate (less than one death per 1,000 cases), transient involvement of the central nervous system as determined by electroencephalographic abnormalities is not unusual [41]. It is generally assumed that in those patients who develop SSPE, measles virus survives in a non-lytic state in brain cells after the initial attack of measles, rather than being reacquired during an asymptomatic second attack of measles.

SSPE has an insidous onset and almost invariably has a fatal outcome. The mean incubation period before the onset of the first neurologic symptoms of SSPE is 6–7 years, with 85% of the cases occurring between 5 and 14 years of age [5, 28–31]. The earliest signs consist of intellectual deterioration and behavioral changes, generally in a grade school student. This is followed by progressive neurologic impairment leading ultimately to cortical blindness, dementia and finally death. Most patients die with deteriorating neurologic status 1–3 years after onset. Rare cases with survival for many years have been reported, as well as fulminant disease leading to death in a matter of months.

III. The Virion

A. Virus Classification

Measles virus belongs to the family Paramyxoviridae [42]. Together with canine distemper and rinderpest viruses, measles is a member of the subgroup, *Morbillivirus* [42]. In addition to being antigenically related [43–47], these three viruses differ from the other paramyxoviruses, such as Sendai, simian virus 5, and Newcastle disease virus, in that they do not adsorb to neuraminic-acid-containing receptors on the surface of cells, nor do they have any neuraminidase activity associated with the virion [48–52]. Unlike canine distemper virus and rinderpest virus, but like the other paramyxoviruses, measles virus has hemagglutination activity [50–58]. Although the host varies, the diseases caused by the morbilliviruses are similar in that infection involves the respiratory tract and results in a cutaneous eruption. In addition, like measles virus in man, canine distemper virus can result in a chronic neurologic disease in the dog.

B. Virion Structure

Measles virions are composed of a nucleocapsid containing a single-stranded RNA genome, surrounded by a lipid-glycoprotein envelope that is derived in part from the host cell membrane. The outer viral envelope is 10–20 nm thick. Short spike-like projections protrude from the surface of the envelope. These spikes consist of the hemagglutination protein H, and the fusion protein F_0 [59–62]. The viral particles are pleomorphic in appearance, although they approximate a spherical shape. Their size ranges from 100 to 300 nm in diameter [5, 52, 63, 64]. Viral particles have a buoyant density of 1.18–1.25 g/cm^3 [59, 65–67]. The virions are fragile and easily disrupted by standard laboratory manipulations used to purify viruses [59, 68]. Thus, they are difficult to isolate.

The internal portion of the virion is the nucleocapsid. The nucleocapsid contains the single-stranded genomic RNA surrounded by viral proteins. Approximately 5% of the nucleocapsid is RNA; the remaining 95% consists of viral proteins [69, 70]. Nucleocapsids have a buoyant density of approximately 1.28–1.30 g/cm³ [59, 69–71]. In the electron microscope nucleocapsids have the appearance of a coiled rod [52, 64, 69, 72, 73]. Thus, the nucleocapsid is often described as being both linear and helical.

The structure of SSPE virions isolated from lytically infected cells is similar to that of measles virions. However, there have been reports that some SSPE virions are more pleomorphic and larger than measles virions and that in infected cells the SSPE nucleocapsids are less well aligned with the cell membrane [74–77]. One study reported SSPE strains that had a buoyant density different from that of wild-type measles virions [78]. In addition, SSPE nucleocapsids appear to be mostly smooth, while wild-type measles nucleocapsids are mostly fuzzy [79]. Similar differences have been reported between nucleocapsids present in the nucleus (smooth) versus nucleocapsids present in the cytoplasm (fuzzy) of cells infected with wild-type measles virus [74, 76, 80–85]. In general, non-productive measles virus infections produce mostly smooth nucleocapsids, while productive measles virus infections produce predominantly fuzzy nucleocapsids. In addition, in productively infected cell cultures containing both smooth and fuzzy nucleocapsids, only the fuzzy nucleocapsids seem to align with the cell membrane and bud out of the cell as mature virions.

C. Measles and SSPE RNA

Up to the present time, isolation and characterization of measles RNA has met with limited success. It is generally agreed that the measles genome consists of a continuous, single strand of RNA with negative sense (i.e., complementary to mRNA). Most measurements of measles genome RNA suggest a sedimentation constant of 50–52S and a molecular weight of 6.2 to 6.4×10^6 [86–89]. Because measles virus is predominately cell-associated, nucleocapsids isolated from infected cells have provided a better source of genomic RNA than have purified virions.

There have been two attempts to define the genetic relatedness of measles and SSPE viruses using hybridization techniques [90, 91]. In one study [90] it was reported that there was only 60% homology between unlabeled 50S SSPE RNA and labeled 18S RNA from measles-virus-infected cells, while there was 100% homology between 18S SSPE RNA and SSPE genome. In a second study [91] it was reported that the SSPE genome contained 10% more information than the measles genome. However, it is difficult to interpret these results, since the purity of the RNA used in the hybridization reactions was not defined in either study. Recent data have shown that cellular mRNA, rRNA, and even DNA may readily contaminate preparations of measles RNA isolated from either infected cells or released virions [*Meissner and Fields,* manuscript in preparation].

Current concepts of measles transcription rely heavily on analogy with other paramyxoviruses, because so little virus-specific RNA can be detected in measles-virus-infected cells. One study [92] compared mRNA from SSPE virus and measles-virus-infected cells on denaturing polyacrylamide gels. The most rapidly migrating SSPE mRNA (the putative M polypeptide message) was reported to migrate more slowly than the corresponding mRNA of measles virus. However, this study is difficult to interpret because the pattern of migration of mRNA from uninfected cells was not presented. Since measles virus infections do not shut off host cell RNA synthesis, it is necessary to demonstrate that each RNA band seen in extracts from infected cells is virus-specific.

One major drawback to the study of measles RNA regulation and expression is the difficulty in isolating large quantities of genome. One recent study [93] attempted to generate complementary DNA using avian myeloblastosis virus reverse transcriptase and measles genomic RNA as template. Alternatively, it is possible to synthesize double-stranded cDNA from viral mRNA extracted from infected cells. Measles mRNA encoding for the nucleocapsid protein (NP), presumably the most abundant viral mRNA in an infected cell, was reported to have been cloned in this fashion [94].

D. Virion Proteins

In general, SSPE virus proteins appear to be similar to measles virus proteins. Measles virions contain seven major proteins: L (large), mol. wt. 200,000; H (hemagglutinin), 80,000; P (phosphoprotein), 70,000; NP (nucleocapsid protein), 60,000; F_0 (fusion), 62,000, composed of polypeptides F_1 and F_2 with molecular weights of 41,000 and 12,000–20,000, respectively; M (matrix), 37,000; and A (actin), 43,000 [59, 65, 68, 95, 96]. With the exception of actin, these proteins are unique and virally encoded. This has been demonstrated by tryptic peptide analysis [97, 99], and immune precipitation of proteins with antibody directed against viral components [97–99].

1. Location of Virion Proteins

Treatment of intact virions with trypsin digests the H and F_0 proteins demonstrating that they are on the viral surface [59, 60]. The L, P, NP, A, and M polypeptides are not digested by trypsin, indicating that they are interior to the cellular membrane. Purification of nucleocapsids from virions reveals that L, P, NP and M are all associated with the nucleocapsid [59]. Preparations of purified membranes from acutely infected cells contain the H and the F_0 proteins as expected, as well as the M protein [59]. The finding that the M protein is strongly associated with both the nucleocapsid and the membrane implies that it is located between the nucleocapsid and the membrane, similar to the matrix protein of other paramyxoviruses.

2. Protein Functions

The large L protein is a minor component of the virion [65, 68]. This protein can be found in nucleocapsid preparations [59] and, by analogy with other negative-strand RNA viruses, it is probably part of the transcription complex.

The H, or hemagglutinin, protein is a surface glycoprotein [59, 61, 62, 68, 100] responsible for the adsorption of measles virus to permissive cells. In addition, the hemagglutinin is the neutralization protein, i.e., it is the main antigen acted upon by neutralizing antibody.

The P protein is a phosphoprotein associated with the nucleocapsid [59, 60, 65, 68, 71]. Although many laboratories report the detection of only minor amounts of P in virions and infected cells, we have demonstrated large amounts of P protein in both infected cells and purified virions [65, 59]. Various factors might result in a reduced recovery of P, since this protein seems to be unusually susceptible to degradation. The function of the P protein has not yet been defined, although by analogy with other paramyxoviruses it is suspected of being part of the transcription complex. In addition, since P is an abundant protein, it may have a structural role in the nucleocapsid.

The NP protein, like P, is a phosphoprotein associated with the nucleocapsid [59, 60, 65, 68, 71]. It is the most abundant viral protein as well as the major structural component of the nucleocapsid and thus, by definition, is the nucleocapsid protein.

F_0 is the fusion protein. It is a glycoprotein composed of two polypeptides, F_1 (41,000 daltons) and F_2 (12,000–20,000 daltons), held together by disulfide bonds [101]. Under the reducing conditions employed in standard SDS-PAGE, F_0 dissociates into F_1 and F_2. The sugar moiety of this glycoprotein resides completely on the F_2 fragment. Since the F_2 fragment is difficult to see on gels, it was originally thought that measles virus had only one glycoprotein [65, 68, 95, 96, 102, 103]. However, under non-reducing conditions, F_0 can be detected on gels, and has been shown to be a glycoprotein [68, 101, 103]. Thus, like the other paramyxoviruses, measles virus has two surface glycoproteins.

The measles virus M protein has long been assigned the function of the matrix protein solely on the basis of analogy with other paramyxoviruses [68, 65]. Recently, the M protein has been shown to be present in preparations of both nucleocapsids and membranes, confirming its role as the matrix protein [59]. One function of the M protein in other negative-strand viruses appears to be that of aligning the nucleocapsid with the cell membrane [see reference 48 for a review]. Such alignment may be one of the final steps in the maturation or budding process of the virus. Defects in this process may be important in the mechanism of measles virus persistence.

In addition to the virally encoded proteins, the measles virion contains another major protein, with a molecular weight of 43,000. Partial proteolytic digestions have recently confirmed that this protein is cellular actin [99]. The role of actin in the measles virus life cycle is currently not known. However, actin appears to be an integral part of the virion rather than simply a surface contaminant, since it is resistant to treatment by trypsin and detergents that remove membrane proteins [59]. Recent investigations in our laboratory suggest that actin may play a role in the budding process [*Stallcup and Fields*, personal commun.]. Chemical inhibitors that interfere with actin filaments seem to decrease the amount of measles virus budding from the cell membrane, although there does not seem to be a reduction in the amount of intracellular measles virus production. It is possible that along with the M protein, actin plays a role in interactions between nucleocapsids and the cell membrane and in interactions between nucleocapsids and the cell cytoskeletal system.

IV. Intracellular Viral Proteins

A. Measles Proteins

Studies of intracellular measles virus protein synthesis, by use of polyacrylamide gel electrophoresis, reveal the same measles virus proteins found in virions [65, 104, 105]. In addition, some minor species are also found that appear to be related to measles virus infection. Two of these proteins have molecular weights of 72,000 and 74,000 and migrate between the H protein and the P protein [65]. A band migrating at about 90,000 may be a precursor of H [59, 65]. Several faint bands are occasionally seen in the region of 54,000 [60, 65, 105]. In addition, a smaller protein with a molecular weight of around 18,000–23,000 has been reported [104, 105]. None of these minor species are routinely detected by immune precipitation [*Wechsler*, unpublished observations], and it is currently not known whether any of these bands represent non-structural viral proteins, whether they are precursors or breakdown products of other viral proteins, or whether they are cellular proteins that are stimulated by measles infection.

B. SSPE Proteins

Several laboratories have studied SSPE proteins in order to compare them to wild-type measles proteins. By most criteria, the SSPE viral proteins are very similar to wild-type measles proteins [106–109]. One difference between the M protein of wild-type measles and some strains of SSPE viruses has been reported. The M protein from SSPE viruses was found to migrate more slowly on SDS-PAGE than the M protein of wild-type measles [106–108]. This implied that the SSPE M protein was larger than the measles M protein. In addition, two cell lines persistently infected with measles virus contained M proteins that migrated more slowly than the M protein of the wild-type virus used to initiate the persistent infections [110]. Thus, it initially appeared that the altered mobility of the M protein might be a biochemical marker for SSPE viruses and that a mutation in the M protein gene might be associated with persistence of measles virus both in tissue culture and in the disease, SSPE.

However, more recent investigations have revealed that occasional wild-type isolates have M proteins with altered mobilities that are similar to those seen in SSPE strains [97–99, 110]. In addition, altered M protein mobilities have not been detected in all of the SSPE strains examined [111, 112]. Thus, although some SSPE virus isolates have M proteins that differ from the common laboratory measles strains, the relationship of aberrant M proteins to SSPE is not absolute.

C. Relative Amounts of Intracellular Viral Proteins

Some SSPE isolates, particularly those that are cell-associated, produce altered amounts of viral proteins compared to wild-type measles [78, 106,

113]. Some strains produce relatively reduced amounts of H or M in comparison to the other viral proteins. Similarly reduced quantities of H and M have been reported for cells persistently infected with measles virus [110]. In addition, analysis of antibody titers to individual measles virus proteins in the sera of patients with SSPE suggests that there is a reduced synthesis of M protein in SSPE brain cells [97–99] (see below). Although it is likely, at the present time there is no conclusive evidence that alterations in the structure, in the relative quantity, or in the stability of viral proteins are involved in the mechanism of measles virus persistence.

V. Viral Replication

A. RNA and Proteins

Measles virus adsorbs to cells, presumably by interaction between the H protein and a cell receptor [100, 114]. Entry probably involves phagocytosis or fusion of the cell and viral membrane, mediated by the fusion protein of the virus. Following entry into the cell, measles virus is uncoated and nucleocapsids are freed.

In comparison to other paramyxoviruses, very little is known about the replication of measles virus RNA. Because the negative-sense genome cannot be used for translation, transcription of mRNA must occur before protein synthesis can begin. Measles virus presumably contains both an RNA-dependent RNA transcriptase and an RNA replicase activity associated with the nucleocapsid [115]. UV transcriptional mapping experiments suggest that mRNA is transcribed beginning at one end of the genome and continuing downstream to the other end [116]. Virus-specific mRNA transcripts have been detected in measles-virus-infected cells [89, 90, 117, 118]. As expected, these mRNAs are complementary to genomic RNA [90, 91].

The time course of synthesis of intracellular measles virus proteins has been examined in cells acutely infected with measles virus [65, 104, 105]. SDS-PAGE revealed the appearance of all viral proteins at approximately the same time post-infection. Thus, there do not seem to be early and late viral proteins in measles-infected cells. Analysis of cells acutely infected with SSPE viruses reveals the same time course of viral protein synthesis [*Wechsler*, unpublished results].

B. Viral Maturation

Studies with the electron microscope (EM) indicate that one of the first visible steps in measles virus infection is the accumulation of inclusion bodies in the cytoplasm of infected cells [80, 83, 84, 87]. These inclusions consist of nucleocapsids. They are present in the perinuclear region of the infected cell and are most often seen in giant multinucleated cells.

Shortly after the appearance of cytoplasmic inclusions, the cell membrane is altered. Spikes appear in the plasma membrane reflecting the insertion of the H and F proteins. Prior to the actual budding process, the nucleocapsids migrate to, and align themselves under, modified areas of the cell membrane [75, 76, 79, 80, 83, 84, 87]. In other paramyxoviruses, this association of nucleocapsids with the cell membrane is known to be mediated by the viral M (matrix) protein [48]. Actin may also be involved in this process, since chemicals that interfere with actin appear to interfere with measles virus nucleocapsid/cell membrane interactions [119]. Finally, areas of the membrane bud out, taking along the nucleocapsid and forming viral particles. This budding process does not appear to be very efficient or exact. EM studies demonstrate many particles containing no nucleocapsids, while other particles contain more than one nucleocapsid.

One of the most striking characteristics of brain cells from patients with SSPE is the abundance of intracellular nucleocapsids. It is also characteristic that these nucleocapsids do not align with the cell membrane and that no budding virus is seen. These non-aligned nucleocapsids have a smooth appearance in electron micrographs as opposed to the fuzzy appearance usually observed in acute, productive infections [79]. The fuzzy and smooth nucleocapsids appear to have different 'surface' protein compositions [71, 79] (possibly different amounts of actin, or the viral matrix protein, M). *Dubois-Dalcq* et al. [79] reported EM studies in which the fuzzy nucleocapsid surface was specifically stained by measles antibody conjugated to an electron-dense material, while smooth nucleocapsids were not stained. In addition, *Robbins* et al. [71] isolated fuzzy and smooth nucleocapsids on cesium chloride gradients and reported that the fuzzy nucleocapsids contained two proteins not found in the smooth nucleocapsids. These two proteins were apparently of cellular origin, and one of them had a molecular weight consistent with that of cellular actin.

The intracellular location of nucleocapsids differs between a productive and a non-productive infection. In productive measles virus infections the majority of the nucleocapsids are in the cytoplasm, while in non-productive infections large quantities of nucleocapsids are found in the cell nucleus [87].

In a productive measles infection alignment of nucleocapsids with the cell membrane appears to be one of the final steps in viral maturation. It is presumed that the M protein mediates the interaction of nucleocapsids and the cell membrane. Thus, it is possible that alterations in the structure or quantity of M protein may play a role in the failure of SSPE nucleocapsids to properly align with the plasma membrane of the infected cell.

C. The Role of the Nucleus in Measles Virus Infections

It has long been recognized that the nucleus of the host cell becomes involved in measles virus infections. EM studies show that late in infection, nucleocapsids are present not only in the cytoplasm, but also in the nuclei [75, 76, 84, 120]. Immunofluorescent studies demonstrate intranuclear staining [1, 81, 121–123], and SDS-PAGE shows the presence of nucleocapsid proteins in the nuclear fraction of infected cells [65]. In addition, the isolation of measles virus RNA from the nuclear fraction of extracts of infected cells has been reported [124, 125]. It is not clear whether the association of virus material with the cell nucleus is an important process in the maturation of measles virus, or whether those viral structures that end up in the nucleus represent an abortive stage in the replicative process.

When cells are infected with undiluted passage stocks of measles virus at high multiplicities of infection, most of the nucleocapsids can be seen in the host cell nucleus rather than in the cytoplasm [87]. In addition, there is a dramatic reduction in the yield of infectious virus particles from cells in which most of the nucleocapsids are in the nucleus. This is similar to what is seen in brain cells from individuals with SSPE [75, 79].

Treatment of infected cells with actinomycin D (which theoretically only affects DNA) may result in a reduction of measles RNA synthesis, measles protein synthesis, and measles virus yield [126, 127; *Meissner and Fields,* in preparation]. In addition, in enucleated cells measles virus yields are reduced almost 100-fold compared to yields in cells containing nuclei, while little or no change is seen in the pattern of viral protein synthesis [128]. However, the enucleation procedure disrupts the cell cytoskeleton. Since the cytoskeletal system is probably involved in measles virus maturation, it is possible that the reduced virus yield seen in enucleated cells may be related to cytoskeletal damage.

As mentioned above, there appears to be a difference between the nucleocapsids located in the nucleus and the nucleocapsids located in the cytoplasm. Cytoplasmic nucleocapsids have a fuzzy granular appearance, while nuclear nucleocapsids appear to be smooth [80, 84]. Similar differences have been reported with SSPE viruses, i.e., in SSPE the predominant nucleocapsid is smooth, while in wild-type measles infection the predominant measles nucleocapsid is fuzzy. It is not known whether the predominantly cytoplasmic location of fuzzy nucleocapsids and the predominantly intranuclear location of smooth nucleocapsids is a result of, or a cause of, their apparent differences.

VI. Growth Characteristics

Wild-type measles virus isolates, that is, virus isolated from patients with natural measles infections, have a wide range of growth characteristics. Plaque morphology, which often reflects the growth patterns of a virus, shows considerable variation amoung wild-type measles isolates. Some isolates make small plaques, others make big plaques. Some isolates make plaques with hazy edges, while others make plaques with discrete edges. The ease with which some wild-type viruses establish persistent infections in tissue culture shows wide variation [*Wechsler,* unpublished observations]. Finally, some wild-type measles isolates are temperature-sensitive compared to the standard laboratory measles virus strains. This wide range of growth characteristics is also found in lytic SSPE virus isolates [21, 106, 107, 112, 117, 129, 130; *Wechsler,* unpublished results]. As a consequence, no specific growth characteristics have been identified that can distinguish between SSPE viruses and measles viruses.

VII. Animal Model Systems

Many investigators have attempted to establish animal model systems for persistent measles and SSPE virus infections [36–40, 131–154]. In hamsters and mice, it has been shown repeatedly that the age of the animal at the time of intracerebral injection is important in the development of encephalitis. Injection of large amounts of wild-type measles virus into the brains of adult hamsters or mice usually results in no apparent symptoms. However, newborn animals show a greater sensitivity to measles virus, usually developing acute encephalitis and finally dying. In addition to acute disease, some infected weanling animals survive and develop a more chronic type of infection.

Several groups have examined the neurovirulence of ts mutants of measles virus. *Haspel* et al. [155, 156] found that the majority of their RNA^- mutants were not virulent when injected into newborn hamster brains, while most of their RNA^+ mutants were virulent. *Burkholtz* et al. [157] found that none of the ts mutants they isolated were virulent in hamsters.

There is evidence that the presence of measles antibodies in newborn animals can significantly alter the outcome of measles virus infection. *Wear and Rapp* [38] studied newborn hamsters that had either high or low levels of maternal measles antibodies. 100% of the animals with low levels of maternal measles antibodies developed encephalitis following intracerebral injections of measles virus. In contrast, only 17% of the newborn hamsters that had high maternal measles antibody levels developed encephalitis following intracerebral injections of measles virus. Of the 83% of the animals with high levels of maternal measles antibodies that survived, 30% developed chronic neural involvement when they were subsequently immunosuppressed with cyclophosphamide. In addition, infectious measles virus could be recovered from the brain tissue of these animals.

Although animal studies using lytic SSPE viruses are similar to animal studies using wild-type measles viruses [140, 146–150, 152–154], several studies with cell-associated SSPE viruses (i.e., cell lines persistently infected with SSPE virus isolates) are of interest [36–40, 78, 150]. These studies suggest that SSPE isolates that have remained cell associated may retain properties that allow them to make SSPE-type infections in some animals. Monkeys injected with the cell-associated SSPE strains Biken or IP-3 [39, 40, 78] develop a persistent infection similar to SSPE. With IP-3-induced disease, the animals developed a high antibody titer against measles in the cerebrospinal fluid as well as intracellular viral inclusions in the brain cells. In addition, chronic encephalitis occurred only in those monkeys with significant measles antibody titers prior to inoculation with the cell-associated SSPE virus. Those animals that did not have prior measles immunity developed acute encephalitis rather than an SSPE-like disease.

The development of SSPE-like disease following intracerebral injections of cell-associated SSPE viruses has also been reported in ferrets [36, 37, 150]. In addition, at least some in vitro measles-virus persistent infections can cause subacute disease in experimental animals similar to that seen with cell-associated SSPE viruses [158]. This is particularly interesting, since it suggests that the ability of measles virus to cause an SSPE-like disease in experimental animals may be related to its ability to cause persistent infections in tissue culture.

VIII. Persistent Infections with Measles Virus

Once infected with wild-type measles virus, permissive cells undergo extensive morphological changes. The cytopathic affect (CPE) involves fusion of cells and migration of the cell nuclei to a central location forming giant multinucleated cells. Eventually, measles virus lyses the infected cells, resulting in cell death. However, measles virus also has the capacity to produce cultures of persistently infected cells.

Classically, persistent infections with measles virus have been initiated by infecting permissive cells at a very low multiplicity of infection (usually less than one plaque-forming unit per 1,000 cells). Following extensive CPE and death of most of the cells, a few cells survive. If these surviving cells are fed at regular intervals (once or twice a week), a monolayer of persistently infected cells may be obtained. These cultures usually go through periods of crisis in which extensive CPE and cell death occurs. This is followed by regrowth of the culture from the surviving cells. Cell lines are often lost during one of these periods of crisis.

A more rapid method of initiating persistent infections with measles virus is to infect cells with virus that has been serially passaged at high multiplicities (undiluted passage). Eventually, this virus will no longer cause lytic cell destruction, and cultures that are persistently infected can be obtained. By analogy with other viruses, defective interfering particles (DIs) are often assumed to be important in this phenomenon [159–162]. Despite considerable effort, little is known about measles virus DIs and their postulated role in measles persistence. In addition, the measles persistent phenotype (nonalignment of nucleocapsids with the membrane and lack of budding) is not that normally associated with interference by DIs [163]. In other viruses, undiluted passage results in a variety of mutants in addition to DIs, and some of these non-DI mutants are capable of interfering with the growth of wild-type virus [164]. Thus, undiluted passage of measles virus may generate mutants other than DI mutants that are involved in the establishment of persistent infections.

The first tissue culture cell lines persistently infected with measles virus were established in the early 1960s [*Rustigian,* 165–167]. Although these are probably the best characterized cell lines persistently infected with measles virus, other laboratories have also established and characterized persistently infected cell lines [168–178]. The viruses in these persistently infected cell lines can often be demonstrated to differ from the virus used to initiate the persistent infection. Thus, some of the viruses recovered from these persistently infected cell lines are temperature-sensitive [110], some contain viral proteins with altered mobilities on SDS-PAGE [110], some have little or no hemagglutination activity [167], and some have little or no M protein [78, 118]. It is not known whether any of these observed viral alterations are involved in the mechanism of measles virus persistence.

Persistently infected cells can be divided into two types [see reference 3 for a review]. Some persistently infected lines (yielder) produce low levels of extracellular virus [165, 166]. Other lines (non-yielder) produce no detectable extracellular virus [167]. Since SSPE brain cells do not release any detectable extracellular virus, the non-yielding persistently infected cell lines resemble the persistent infections in brain cells of individuals with SSPE more closely than do the yielding persistently infected cell lines. In fact, cell-associated SSPE viruses that are recovered as persistently infected cell lines are usually of the non-yielding type [22, 33–35].

The role of the host cell in measles virus persistence may be as important as that of the virus. In hamster cells persistently infected with measles virus, virus can be 'rescued' within 6 h of co-cultivation with BSC-1 cells, even in the absence of new viral RNA or protein synthesis [171, 176, 177]. During the co-cultivation, the BSC-1 cells fuse with the persistently infected cells due to the presence of the fusion protein in the membrane of the persistently infected cell. This fusion allows the mixing of cellular proteins and possibly cell cytoskeletal structures between the cells. The rapid 'rescue' of infectious virus without new RNA or protein synthesis suggests that viral maturation in the persistently infected cells was blocked at a very late stage and, in addition, that the block was due to a cellular rather than a viral function.

More direct evidence for the importance of the cell in measles virus persistence is the fact that some viral stocks are capable of rapidly establishing a persistent infection in one cell type but not another. For example, virus recovered from a persistently infected HeLa cell line (K11) immediately produces a persistent infection in normal HeLa cells, yet it causes an acute lytic infection in CV-1 or Vero cells [110]. Measles virus grown in Vero cells rapidly causes a persistent infection in BGM cells even though the same viral stock is lytic in Vero cells [179]. In MDBK cells, L cells and mouse neuron cultures, wild-type measles virus may immediately establish a persistent infection with no CPE, but with continuous low levels of virus production [180, 181; *Wechsler*, unpublished observations]. Thus, the cell as well as the virus plays an important role in persistent infections by measles virus.

IX. Immunology

A. Neutralization

It has been reported that there is a difference between the neutralization characteristics of SSPE viruses and wild-type viruses. Specifically it was found that some SSPE viruses were more resistant to neutralization than wild-type viruses, regardless of the antisera employed in the neutralization test [182]. However, other reports [183, 184] have not noted any difference in the neutralization properties of SSPE and wild-type measles isolates.

B. Antigenic Cross-Reactivity

Hall and ter Meulen [92] reported that the M protein of one SSPE virus strain is antigenically distinct from the M protein of one wild-type measles virus strain. Due to the difficulty in purifying M proteins, control studies were not done to compare M proteins of different wild-type measles isolates. Although this initial report has not been confirmed [98, 99], it nevertheless led to several studies on the nature of the host's antibody response to wild-type and SSPE viruses.

C. Host Humoral Immune Response

It has been well documented that patients with SSPE have high anti-measles antibody titers [see references 3 and 5 for reviews]. The degree and the persistence of the antibody response presumably reflects the continued exposure of the immune system to measles virus. The antibody response to individual measles proteins has been examined by immune precipitation of radioactively labeled viral proteins followed by SDS-PAGE [97–99]. These studies have demonstrated that, on the average, sera from patients with SSPE have a reduced antibody response to the M protein of measles virus compared both to sera from adults who experienced natural measles infections in childhood and to sera from individuals 3–5 weeks after natural measles infection. This decreased antibody response to the M protein in SSPE sera occurs despite a vigorous antibody response to the other viral proteins. It should be emphasized that this reduced antibody response against the M protein is a statistical phenomenon rather than an 'all or nothing' situation. Thus, some SSPE sera are capable of precipitating large quantities of M protein, while some convalescent sera precipitate none. However, when numerous sera are examined, there is a statistically significant difference between the amount of M protein precipitated by SSPE sera compared to the amount precipitated by normal or convalescent sera ($p < 0.003$) [98, 99].

It is unlikely that this diminished antibody response to the M protein represents a defect in the immune response of the host, since there are large amounts of antibody produced to the other viral proteins. The inability of SSPE sera to precipitate M protein is not due to antibody excess, because dilutions of SSPE sera do not precipitate increased amounts of M protein [98]. In addition, it is unlikely that there is an inhibiting factor in SSPE sera that prevents precipitation of M, since mixtures of SSPE sera and sera containing anti-M antibodies are capable of precipitating M proteins [97].

The diminished antibody response to the M protein of measles virus in SSPE does not appear to be due to antigenic differences between the M proteins of wild-type measles and SSPE viruses. Sera from individuals with SSPE precipitated less M protein regardless of whether the M protein tested was derived from wild-type measles viruses or from SSPE viruses. In addition, sera from normal adults with a history of measles infection in childhood and sera from individuals convalescing from measles precipitated M protein from both wild-type and SSPE viruses to an equal extent [98, 99]. In contrast to the initial report that wild-type measles and SSPE viruses have antigenically distinct M proteins [92], antisera prepared in rabbits against five SSPE strains precipitated M proteins from both wild-type and SSPE viruses, as did antisera prepared in rabbits against two wild-type measles strains [98, 99]. Therefore, the reduced amount of anti-M antibody in SSPE sera suggests that in SSPE patients only small amounts of the M protein are present and that these low levels of M protein account for the minimal host response to this protein [97–99, 109].

The hypothesis of reduced M protein in individuals with SSPE has been supported by further evidence. In a strain of HeLa cells persistently infected with measles virus, very little M protein is synthesized [110]. Preliminary experiments suggest that guinea pigs injected with this persistently infected cell line make good antibody response to all of the proteins except the M protein [*Wechsler,* unpublished results]. This may be similar to what is occurring in individuals with SSPE. *Hall and Choppin* [109] have reported that brain cells isolated from one individual with SSPE contain all of the measles proteins except the M protein, thus supporting the hypothesis that less M is present in individuals with SSPE. In addition, two cell-associated SSPE strains,

IP-3 and Biken, that were recovered as persistent infections, appear to produce reduced amounts of M protein [78], again suggesting that in the SSPE brain cells only small amounts of M protein are present. In the same study it was reported that these isolates produce normal amounts of mRNA for all of the viral proteins, including the M protein. This suggests that the defect in the amount of M protein present in these cells is not at the level of transcription of mRNA. Thus it is possible that the defect is at the translation step, or it may be that M protein synthesized in these cells is rapidly broken down.

It should be noted that despite the evidence presented above, there have been reports that disagree with the conclusion that SSPE sera tend to have reduced antibody levels to the M protein [185–187].

X. Mechanism(s) of Measles Virus Persistence

Many theories have been proposed to explain the mechanism(s) for persistent infections with measles virus. These theories include defective interfering particles, temperature-sensitive mutants, alterations in the amount of specific viral proteins, alterations in the primary structure of virus proteins, interferon, antibody modulation [188, 189] and host cell factors [see reference 4 for a review]. Because of the relative ease of examining viral rather than cellular changes, research on the mechanism(s) of persistent infections with measles virus has generally focused on alterations of the virus, rather than the cell. However, it is becoming clear that the cell, as well as the virus, plays an important role in the ability of measles virus to cause a persistent infection.

Any theory attempting to explain the mechanism(s) of measles virus persistence should also explain the measles persistent phenotype (non-alignment of nucleocapsids with the cell membrane, and reduced budding). There are two specific stages during measles virus maturation in which alterations might readily be envisioned to result in both persistent infections and the persistent phenotype. These stages are (1) viral nucleocapsid/cell cytoskeleton interactions, that presumably are responsible for movement of nucleocapsids from the perinuclear region to the cell membrane; and (2) interactions between nucleocapsids and the cell membrane. Nucleocapsid/cytoskeleton interactions probably involve one or more of the viral nucleocapsid-associated proteins (M, NP, or P) and one or more cell cytoskeleton proteins (i.e., actin). Nucleocapsid/membrane interactions are mediated by M protein, possibly in conjunction with one or more additional viral proteins (nucleocapsid or membrane) or cellular proteins (membrane or cytoskeletal). Thus, differences in the virus or the cell (as a result of mutation, variations between viral isolates, or variations between cell types), may be involved in persistent infections with measles virus.

XI. Conclusions

At the present time, no differences have been found between lytic SSPE viruses and lytic measles viruses that might explain the mechanism of measles virus persistence. In addition, there are no clear distinctions between cell-associated SSPE viruses and cell-associated persistent measles infections in vitro. Instead, the important differences appear to be between productive, lytic viruses and non-productive, cell-associated viruses, regardless of whether they are SSPE viruses or measles viruses.

Since cells persistently infected with measles virus generally contain large quantities of viral nucleocapsids, it is likely that the mechanism(s) of measles virus persistence is at a late stage in viral maturation. In addition to the virus, the cell plays an important role in persistent infections. In particular, virus/cell interactions, involving movement of viral nucleocapsids to the cell membrane and alignment of these nucleocapsids with the cell membrane, are potential steps at which abnormalities might result in measles virus persistence. Hopefully, continued investigations into the differences between lytic and persistent infections with measles and SSPE viruses will lead to a better understanding not only of the pathogenesis of SSPE but also other slow virus diseases.

References

1 Enders, J.F.; Peebles, T.C.: Propagation in tissue cultures of cytopathogenic agents from patients with measles. Proc. Soc. exp. Biol. Med. *86:* 277–286 (1954).

2 Enders, J.F.; Katz, S.; Milovanovic, M.C.; Halloway, A.: Studies on an attenuated measles-virus vaccine. I. Development and preparation of the vaccine: techniques for assay of effects of vaccination. New Engl. J. Med. *263:* 153–169 (1960).

3 Morgan, E.M.; Rapp, F.: Measles virus and its associated diseases. Bact. Rev. *41:* 636–666 (1977).

4 Rima, B.K.; Martin, S.J.: Persistent infection of tissue culture cells by RNA viruses. Med. Microbiol. Immunol. *162:* 89–118 (1977).

5 ter Meulen, V.; Hall, W.W.: Slow virus infections of the nervous system: virological, immunological and pathogenetic considerations. J. gen. Virol. *41:* 1–25 (1978).

6 ter Meulen, V.; Katz, M.; Muller, D.: Subacute sclerosing panencephalitis: a review. Curr. Top. Microbiol. Immunol. *57:* 1–38 (1972).

7 Bouteille, M.; Fountaine, C.; Vedrenne, C.; Delarue, J.: Sur un cas d'encéphalite subaigue à inclusions. Etude anatomoclinique et ultrastructurale. Revue neurol. *113:* 454–458 (1965).

8 Dayan, A.D.; Gostling, J.V.T.; Greaves, J.L.; Stevens, D.W.; Woodhouse, M.A.: Evidence of a pseudomyxovirus in the brain in subacute sclerosing leucoencephalitis. Lancet *i:* 980–981 (1967).

9 Freeman, J.M.; Magoffin, R.L.; Lennette, E.H.; Herndon, R.M.: Additional evidence of the relation between subacute inclusion body encephalitis and measles virus. Lancet *ii:* 129–131 (1967).

10 Herndon, R.M.; Rubinstein, L.J.: Light and electron microscopy observations in the development of viral particles in the inclusions of Dawson's encephalitis (subacute sclerosing panencephalitis). Neurology, Minneap. *18:* 8–20 (1967).

11 Katz, M.; Oyanagi, S.; Koprowski, H.: Subacute sclerosing panencephalitis: structures resembling myxovirus nucleocapsid in cells cultured from brains. Nature, Lond. *222:* 888–899 (1969).

12 Perier, O.; Thiry, L.; Vanderhaeghen, J.J.; Pelc, S.: Attempts at experimental transmission and electron microscopic observations in subacute sclerosing panencephalitis. Neurology, Minneap. *18:* 138–143 (1968).

13 Shaw, C.M.; Buchan, G.C.; Carlson, C.B.: Myxovirus as a possible etiologic agent in subacute inclusion-body encephalitis. New Engl. J. Med. *277:* 511–515 (1967).

14 Tellez-Nagel, J.; Harter, D.H.: Subacute sclerosing leucoencephalitis. I. Clinical, pathological, electron microscopic and virological observations. J. Neuropath. exp. Neurol. *25:* 560–581 (1966).

15 Tellez-Nagel, J.; Harter, D.H.: Subacute sclerosing leucoencephalitis: ultrastructure of intranuclear and intracytoplasmic inclusions. Science *154:* 899–901 (1966).

16 Connolly, J.H.; Allen, I.V.; Hurwitz, L.J.; Miller, J.H.D.: Measles-virus antibody and antigen in subacute sclerosing panencephalitis. Lancet *i:* 542–544 (1967).

17 Horta-Barbosa, L.; Fuccillo, D.A.; London, W.T.; Jabbour, J.T.; Zeman, W.; Sever, J.L.: Isolation of measles virus from brain cell cultures of two patients with subacute sclerosing panencephalitis. Proc. Soc. exp. Biol. Med. *132:* 272–277 (1969).

18 Horta-Barbosa, L.; Fuccillo, D.A.; Sever, J.L.; Zeman, W.: Subacute sclerosing panencephalitis: isolation of measles virus from a brain biopsy. Nature, Lond. *221:* 974 (1969).

19 Chen, T.T.; Watanabe, I.; Zeman, W.; Mealey, J., Jr.: Subacute sclerosing panencephalitis: propagation of measles virus from brain biopsy in tissue culture. Science *163:* 1193–1194 (1969).

20 Payne, F.E.; Baublis, V.V.; Itabashi, H.H.: Isolation of measles virus from cell cultures of brain from a patient with subacute sclerosing panencephalitis. New Engl. J. Med. *281:* 585–589 (1969).

21 ter Meulen, V.; Katz, M.; Kackell, Y.F.; Barbanti-Brodano, G.; Koprowski, H.; Lennette, E.H.: Subacute sclerosing panencephalitis: in vitro characterization of viruses isolated from brain cells in culture. J. infect. Dis. *126:* 11–17 (1972).

22 Ueda, S.; Okuno, Y.; Hamamoto, Y.; Ohya, H.: Subacute sclerosing panencephalitis (SSPE): isolation of a defective variant of measles virus from brain obtained at autopsy. Biken's J. *18:* 113–122 (1975).

23 Horta-Barbosa, L.; Hamilton, R.; Wittig, B.; Fuccillo, D.; Sever, J.L.: Subacute sclerosing panencephalitis: isolation of suppressed measles virus from lymph node biopsies. Science *173:* 840–841 (1971).

24 Parker, J.C.; Klintworth, G.K.; Graham, D.G.; Griffith, J.F.: Uncommon morphologic features in subacute sclerosing panencephalitis (SSPE). Am. J. Path. *61:* 275–284 (1970).

25 Johannes, R.S.; Sever, J.L.: Subacute sclerosing panencephalitis. A. Rev. Med. *26:* 589–601 (1975).

26 Johnson, R.T.: Subacute sclerosing panencephalitis. J. infect. Dis. *121:* 227–230 (1970).

27 Dawson, J.R.: Cellular inclusions in cerebral lesions of lethargic encephalitis. Am. J. Path. *9:* 7–16 (1933).

28 Modlin, J.F.; Jabbour, J.T.; Witte, J.J.; Halsey, N.A.: Epidemiologic studies of measles, measles vaccine, and subacute sclerosing panencephalitis, Pediatrics, Springfield *59:* 505–512 (1977).

29 Modlin, J.F.; Halsey, N.A.; Eddins, D.L.; Conrad, J.L.; Jabbour, J.T.; Chien, L.; Robinson, H.: Epidemiology of subacute sclerosing panencephalitis. J. Pediat. *94:* 231–236 (1979).

30 Jabbour, J.T.; Garcia, J.G.; Lemmi, H.; Ragland, J.; Duenas, D.A.; Sever, J.L.: Subacute sclerosing panencephalitis. A multidisciplinary study of eight cases. J. Am. med. Ass. *207:* 2248–2254 (1972).

31 Sever, J.L.; Ellenberg, J.H.; Krebs, H.M.; Jabbour, J.T.: Subacute sclerosing panencephalitis and measles vaccine. Pediat. Res. *8:* 114 (1974).

32 Katz, M.; Koprowski, H.: The significance of failure to isolate infectious viruses in cases of subacute sclerosing panencephalitis. Arch. ges. Virusforsch. *41:* 390–393 (1973).

33 Burnstein, T.; Jacobsen, L.B.; Zeman, W.; Chen, T.T.: Persistent infection of BSC-1 cells by defective measles virus derived from subacute sclerosing panencephalitis. Infect. Immunity *10:* 1378–1382 (1974).

34 Doi, Y.; Sampe, T.; Nakajima, M.; Okawa, S., Katog, T.; Itoh, H.; Sato, T.; Oguchi, K ; Kumanishi, T.; Tsubaki, T.: Properties of a cytopathic agent isolated from a patient with subacute sclerosing panencephalitis in Japan. Jap. J. med. Sci. Biol. *25:* 321–333 (1972).

35 Dubois-Dalcq, M.; Horta-Barbosa, L.; Hamilton, R.; Sever, J.L.: Comparison between productive and latent SSPE viral infection in vitro. Lab. Invest. *30:* 241–250 (1974).

36 Thormar, H.; Jervis, G.A.; Karl, S.C.; Brown, H.R.: Passage in ferrets of encephalitogenic cell-associated measles virus isolated from brain of a patient with subacute sclerosing panencephalitis. J. infect. Dis. *127:* 678–685 (1973).

37 Thormar, H.; Mehta, P.D.; Brown, H.R.: Comparison of wild-type and subacute sclerosing panencephalitis strains of measles virus. J. exp. Med. *148:* 674–691 (1978).

38 Wear, D.J.; Rapp, F.: Latent measles virus infection of the hamster central nervous system. J. Immun. *107:* 1593–1598 (1971).

39 Albrecht, P.; Burnstein, T.; Kutch, M.J.; Hicks, J.T.; Ennis, F.A.: Subacute sclerosing panencephalitis: experimental infection in primates. Science *195:* 64–66 (1977).

40 Ueda, S.; Otsuka, T.; Okuno, Y.: Experimental subacute sclerosing panencephalitis (SSPE) induced in a monkey by subcutaneous inoculation with a defective SSPE virus. Biken's J. *18:* 179–181 (1975).

41 Gibbs, F.A.; Gibbs, E.L.; Carpenter, P.R.; Spies, H.W.: Electroencephalographic changes in 'uncomplicated' childhood disease. J. Am. med. Ass. *171:* 1050–1055 (1959).

42 Fenner, F.: Classification and nomenclature of viruses. Second report of the International Committee on Taxonomy of Viruses. Intervirology *7:* 1–116 (1976).

43 Gillespie, J.H.; Karzon, D.T.: A study of the relationship between canine distemper and measles in the dog. Proc. Soc. exp. Biol. Med. *105:* 547–551 (1960).

44 Imagawa, D.T.: Relationships among measles, canine distemper and rinderpest viruses. Prog. med. Virol., vol. 10, pp. 160–193 (Karger, Basel 1968).

45 Orvell, C.; Norrby, E.: Further studies on the immunologic relationships among measles, distemper, and rinderpest viruses. J. Immun. *113:* 1850–1858 (1974).

46 Waterson, A.P.; Rott, R.; Ruckle-Enders, G.: The components of measles virus and their relation to rinderpest and distemper. Z. Naturf. *18b:* 377–384 (1963).

47 Yamanouchi, K.; Kobune, F.; Fukuda, A.; Hayami, M.; Shishido, A.: Comparative immunofluorescent studies on measles, canine distemper and rinderpest viruses. Arch. ges. Virusforsch. *29:* 90–100 (1970).

48 Choppin, P.W.; Compans, R.W.: Reproduction of paramyxoviruses; in Fraenkel-Conrat, Wagner, Comprehensive virology; vol. 4, pp. 95–178 (Plenum Press, New York 1974).

49 Howe, C.; Schluederberg, A.: Neuraminidase associated with measles virus. Biochem. biophys. Res. Commun. *40:* 606–607 (1970).

50 Norrby, E.: Hemagglutination by measles virus. II. Properties of the hemagglutinin and of the receptors on the erythrocytes. Arch. ges. Virusforsch. *12:* 164–172 (1962).

51 Peries, J.R.; Chany, C.: Studies on measles viral hemagglutination. Proc. Soc. exp. Biol. Med. *110:* 477–482 (1962).

52 Waterson, A.P.: Measles virus. Arch. ges. Virusforsch. *16:* 57–80 (1965).

53 Gould, E.A.; Cosby, S.L.; Shirodaria, P.V.: Salt-dependent hemagglutinating measles virus in SSPE. J. gen. Virol. *33:* 139–142 (1976).

54 Norrby, E.: Hemagglutination by measles virus. 1. The production of hemagglutinin in tissue culture and the influence of different conditions on the hemagglutinating system. Arch. ges. Virusforsch. *12:* 153–163 (1962.

55 Norrby, E.: Hemagglutination by measles virus. 4. A simple procedure for production of high potency antigen for hemagglutination-inhibition (HI) tests. Proc. Soc. exp. Biol. Med. *8:* 814–818 (1962).

56 Norrby, E.; Gollmar, Y.: Identification of measles virus-specific hemolysis-inhibiting antibodies separate from hemagglutination-inhibiting antibodies. Infect. Immunity *11:* 231–239 (1975).

57 Peries, J.R.; Chany, C.: Activité hémagglutinante et hémolytique du virus morbilleux. C.r. hebd. Séanc. Acad. Sci., Paris *251:* 820–821 (1960).

58 Rosen, L.: Hemagglutination and hemagglutination-inhibition with measles virus. Virology *13:* 139–141 (1961).

59 Stallcup, K.C.; Wechsler, S.L.; Fields, B.N.: Purification of measles virus and characterization of subviral components. J. Virol. *30:* 166–176 (1979).

60 Tyrrell, D.L.J.; Norrby, E.: Structural polypeptides of measles virus. J. gen. Virol. *39:* 219–229 (1978).

61 Anttonen, O.; Jokinen, M.; Salmi, A.; Vainionpaa, R.; Gahmberg, C.G.: The glycoprotein of measles virus. Biochem. J. *185:* 189–194 (1979).

62 Wild, F.; Greenland, T.: A study of the measles virus-induced proteins incorporated into the cell membrane. Intervirology *11:* 275–281 (1979).

63 Mountcastle, W.E.; Compans, R.W.; Caliguiri, L.A.; Choppin, P.W.: Nucleocapsid protein subunits of simian virus 5, Newcastle disease virus, and Sendai virus. J. Virol. *6:* 677–684 (1970).

64 Waterson, A.P.; Cruickshank, J.G.; Lawrence, G.D.; Kanarek, A.D.: The nature of measles virus. Virology *15:* 379–382 (1961).

65 Wechsler, S.L.; Fields, B.N.: Intracellular synthesis of measles virus-specified polypeptides. J. Virol. *25:* 285–297 (1978).

66 Norrby, E.C.J.; Magnusson, P.; Falksveden, L.G.; Gronberg, M.: Separation of measles virus components by equilibrium centrifugation in CsCl gradients. II. Studies on the large and the small hemagglutinin. Arch. ges. Virusforsch. *14:* 462–473 (1964).

67 Numazaki, Y.; Karzon, D.T.: Density separable fractions during growth of measles virus. J. Immun. *97:* 458–469 (1966).

68 Mountcastle, W.E.; Choppin, P.W.: A comparison of the polypeptides of four measles virus strains. Virology *78:* 463–474 (1977).

69 Waters, D.J.; Bussell, R.H.: Isolation and comparative study of the nucleocapsids of measles and canine distemper viruses from infected cells. Virology *61:* 64–79 (1973).

70 Waters, D.J.; Hersh, R.T.; Bussell, R.H.: Isolation and characterization of measles nucleocapsid from infected cells. Virology *48:* 278–281 (1972).

71 Robbins, S.J.; Bussell, R.H.; Rapp, F.: Isolation and partial characterization of two forms of cytoplasmic nucleocapsids from measles virus-infected cells. J. gen. Virol. *47:* 301–310 (1979).

72 Norrby, E.C.J.; Magnusson, P.: Some morphological characteristics of the internal component of measles virus. Arch. ges. Virusforsch. *17:* 443–447 (1965).

73 Norrby, E.; Hammarskjold, B.: Structural components of measles virus. Microbios *5:* 17–29 (1972).

74 Raine, C.S.; Feldman, L.A.; Sheppard, R.D.; Barbosa, L.H.; Bornstein, M.B.: Subacute sclerosing panencephalitis virus. Observations on a neuroadapted and non-neuroadapted strain in organotypic central nervous system cultures. Lab. Invest. *31:* 42–53 (1974).

75 Dubois-Dalcq, M.: Pathology of measles virus infection of the nervous system: comparison with multiple sclerosis. Int. Rev. exp. Path. *19:* 101–135 (1979).

76 Raine, C.S.; Feldman, L.A.; Sheppard, R.D.; Bornstein, M.B.: Ultrastructure of measles virus in cultures of hamster cerebellum. J. Virol. *4:* 169–181 (1969).

77 Fleury H.; Bonnez, W.; Pometan, J.P.; Dupasquier, P.: Differences in early ultrastructural aspects of the replication of measles and subacute sclerosing panencephalitis viruses in a cell culture from a human astrocytoma. J. Neuropath. exp. Neurol. *39:* 131–137 (1980).

78 Milstien, J.B.; Albrecht, P.; Seifried, A.S.: Studies on cell-associated SSPE isolates: the role of matrix protein; in Abstr. First Ames-Yissum Virology Conf. on Latent and Persistent Virus Infections in Man, Israel 1979.

79 Dubois-Dalcq, M.; Barbosa, L.H.; Hamilton, R.; Sever, J.L.: Comparison between productive and latent subacute sclerosing panencephalitis viral infection in vitro. Lab. Invest. *30:* 241–250 (1974).

80 Dubois-Dalcq, M.; Barbosa, L.H.: Immunoperoxidase stain of measles antigen in tissue culture. J. Virol. *12:* 909–918 (1973).

81 Oyanagi, S.; ter Meulen, V.; Muller, D.; Katz, M.; Koprowski, H.: Electron microscopic observations in subacute sclerosing panencephalitis brain cell cultures: their correlation with cytochemical and immunocytological findings. J. Virol. *6:* 370–379 (1970).

82 Feldman, L.A.; Raine, C.S.; Sheppard, R.D.; Bornstein, B.: Virus-host cell relationships in measles-infected cultures of central nervous tissue. J. Neuropath. exp. Neurol. *31:* 624 (1972).

83 Nakai, M.; Imagawa, D.T.: Electron microscopy of measles virus replication. J. Virol. *3:* 187–197 (1969).

84 Oyanagi, S.; ter Meulen, V.; Katz, M.; Koprowski, H.: Comparison of subacute sclerosing panencephalitis and measles virus: an electron microscope study. J. Virol. *7:* 176–187 (1971).

85 Raine, C.S.; Feldman, L.A.; Sheppard, R.D.; Bornstein, M.B.: Ultrastructural study of long-term measles infection in cultures of hamster dorsal-root ganglion. J. Virol. *8:* 318–329 (1971).

86 Hammarskjold, B.; Norrby, E.: A comparison between virion RNA of measles virus and some other paramyxoviruses. Med. Microbiol. Immunol. *160:* 99–104 (1974).

87 Nakai, T.; Shand, F.L.; Howatson, A.F.: Development of measles virus in vitro. Virology *38:* 50–67 (1969).

88 Schluederberg, A.: Measles virus RNA. Biochem. biophys. Res. Commun. *42:* 1012–1015 (1971).

89 Winston, S.H.; Rustigian, R.; Bratt, M.A.: Persistent infection of cells in culture by measles virus. III. Comparison of virus-specific RNA synthesized in primary and persistent infection in HeLa cells J. Virol. *11:* 926–932 (1973).

90 Yeh, J.: Characterization of virus-specific RNAs from subacute sclerosing panencephalitis virus-infected CV-1 cells. J. Virol. *12:* 962–968 (1973).

91 Hall, W.W.; ter Meulen, V.: RNA homology between subacute sclerosing panencephalitis and measles viruses. Nature, Lond. *264:* 474–477 (1976).

92 Hall, W.W.; ter Meulen, V.: Membrane proteins of subacute sclerosing panencephalitis and measles virus. Nature, Lond. *272:* 460 (1978).

93 Coulter-Mackie, M.B.; Bradbury, W.C.; Dales, S.; Flintoff, W.F.; Morris, V.L.: *In vivo* and *in vitro* models of demyelinating diseases. IV. Isolation of Halle measles virus-specific RNA from BGMK cells and preparation of complementary DNA. Virology *102:* 327–338 (1980).

94 Gorecki, M.; Rozenblatt, S.: Cloning of DNA complementary to the measles virus mRNA encoding nucleocapsid protein. Proc. natn. Acad. Sci. USA *77:* 3686 (1980).

95 Bussell, R.H.; Waters, D.J.; Seals, M.K.; Robinson, W.S.: Measles, canine distemper and respiratory syncytial virions and nucleocapsids. A comparative study of their structure, polypeptide and nucleic acid composition. Med. Microbiol. Immunol. *160:* 105–124 (1974).

96 Waters, D.J.; Bussell, R.H.: Polypeptide composition of measles and canine distemper viruses. Virology *55:* 554–557 (1973).

97 Hall, W.W.; Lamb, R.A.; Choppin, P.W.: Measles and subacute sclerosing panencephalitis virus proteins: lack of antibody to the M protein in patients with SSPE. Proc. natn. Acad. Sci. USA *76:* 2047–2051 (1979).

98 Wechsler, S.L.; Weiner, H.L.; Fields, B.N.: Immune response in subacute sclerosing panencephalitis: reduced antibody response to the matrix protein of measles virus. J. Immun. *123:* 884–889 (1979).

99 Wechsler, S.L.; Meissner, H.C.; Ray, U.R.; Weiner, H.L.; Rustigian, R.; Fields, B.N.: Immune response in subacute sclerosing panencephalitis and multiple sclerosis: antibody response to measles virus proteins; in Bishop, Compans, Replication of negative strand viruses (Elsevier/North Holland, Amsterdam, in press, 1981).

100 Fenger, T.; Howe, C.: Isolation and characterization of erythrocyte receptors for measles virus. Proc. Soc. exp. Biol. Med. *162:* 299–303 (1979).

101 Hardwick, J.M.; Bussell, R.H.: Glycoproteins of measles virus under reducing and non-reducing conditions. J. Virol. *25:* 687–692 (1978).
102 Breschkin, A.M.; Walmer, B.; Rapp, F.: Hemagglutination variant of measles virus. Virology *80:* 441–444 (1977).
103 Hardwick, J.M.; Bussell, R.H.: Glycoproteins of measles virus under reduced and non-reduced conditions. Abstr. Annu. Meet. Am. Soc. Microbiol. p. 232 (Abstr. No. 166) (1976).
104 Graves, M.S.; Silver, S.M.; Choppin, P.W.: Measles virus polypeptide synthesis in infected cells. Virology *86:* 254–263 (1978).
105 Vainionpaa, R.; Ziola, B.; Salmi, A.: Measles virus polypeptides in purified virions and in infected cells. Acta path. microbiol. scand. *86:* 379–385 (1978).
106 Wechsler, S.L.; Fields, B.N.: Differences between the intracellular polypeptides of measles and subacute sclerosing panencephalitis virus. Nature, Lond. *272:* 458–460 (1978).
107 Wechsler, S.L.; Stallcup, K.C.; Fields, B.N.: A comparison of the intracellular polypeptides of measles and subacute sclerosing panencephalitis virus; in Mahy, Barry, Negative strand viruses and the host cell, pp. 169–180 (Academic Press, New York 1978).
108 Schluederberg, A.; Chavanich, S,; Lipman, M.B.; Carter, C.: Comparative molecular weight estimates of measles and subacute sclerosing panencephalitis virus structural polypeptides by simultaneous electrophoresis in acrylamide slab gels. Biochem. biophys. Res. Commun. *58:* 647 (1974).
109 Hall, W.W.; Choppin, P.W.: Evidence for lack of synthesis of the M polypeptide of measles virus in brain cells in subacute sclerosing panencephalitis. Virology *99:* 443–447 (1979).
110 Wechsler, S.L.; Rustigian, R.; Stallcup, K.C.; Byers, K.B.; Winston, S.H.; Fields, B.N.: Measles virus-specified polypeptide synthesis in two persistently infected HeLa cell lines. J. Virol. *31:* 677–684 (1979).
111 Rima, B.K.; Martin, S.J.; Gould, E.A.: A comparison of polypeptides in measles and SSPE virus strains. J. gen. Virol. *42:* 603–608 (1979).
112 Miller, C.A.: Intranuclear polypeptides of measles and subacute sclerosing panencephalitis virus-infected cultures. Virology *101:* 272–276 (1980).
113 Machamer, C.E.; Hayes, E.C.; Zweerink, H.J.: Cell-associated SSPE virus does not express a matrix protein. Abstr. Annu. Meet. Am. Soc. Microbiol., p. 261 (1979).
114 Duffy, P.D.; Howe, C.: Rhesus erythrocyte glycoprotein receptor for measles virus. Abstr. Annu. Meet. Am. Soc. Microbiol., p. 232 (Abstr. No. 167) (1976).
115 Seifried, A.S.; Albrecht, P.; Milstien, J.B.: Characterization of an RNA-dependent RNA polymerase activity associated with measles virus. J. Virol. *25:* 781–787 (1978).
116 Hall, W.W.; Kiessling, W.R.; ter Meulen, V.: Biochemical comparison of measles and subacute sclerosing panencephalitis (SSPE) viruses; in Mahy, Barry, Negative strand viruses and the host cell, pp. 143–156 (Academic Press, New York 1978).
117 Carter, C.; Schluederberg, A.; Black, F.L.: Viral RNA synthesis in measles virus-infected cells. Virology *53:* 379–383 (1973).
118 Kiley, M.P.; Payne, F.E.: Replication of measles virus: continued synthesis of nucleocapsid RNA and increased synthesis of mRNA in the presence of cycloheximide. J. Virol. *14:* 758–764 (1974).
119 Tyrrell, D.L.J.; Ehrnst, A.: Transmembrane communication in cells chronically infected with measles virus. J. Cell Biol. *81:* 396–402 (1979).
120 Baker, R.F.; Gordon, I.; Rapp, F.: Electron-dense crystallites in nuclei of human amnion cells infected with measles virus. Nature, Lond. *185:* 790–791 (1960).

121 Llanes-Rodas, R.; Liu, C.: A study of measles virus infection in tissue culture cells with particular reference to the development of intranuclear inclusion bodies. J. Immun. *95:* 840–845 (1965).

122 Norrby, E.: Intracellular accumulation of measles virus nucleocapsid and envelope antigens. Microbios. *5:* 31–40 (1972).

123 Rapp, F.; Gordon, I.; Baker, R.F.: Observations of measles virus infection of cultured human cells. I. A study of development and spread of virus antigen by means of immunofluorescence. J. biophys. biochem. Cytol. *7:* 43–48 (1960).

124 Carter, C.; Black, F.L.; Schluederberg, A.: Nucleus-associated RNA in measles virus-infected cells. Biochem. biophys. Res. Commun. *54:* 411–416 (1973).

125 Schluederberg, A.; Chavanich, S.: The role of the nucleus in measles virus replication. Med. Microbiol. Immunol. *160:* 85–90 (1974).

126 Hall, W.W.; ter Meulen, V.: The effects of actinomycin D on RNA synthesis in measles virus-infected cells. J. gen. Virol. *34:* 391–396 (1977).

127 Schluederberg, A.; Williams, C.A.; Black, F.L.: Inhibition of measles virus replication and RNA synthesis by actinomycin D. Biochem. biophys. Res. Commun. *48:* 657–661 (1972).

128 Follett, E.A.C.; Pringle, C.R.; Pennington, T.H.; Shirodaria, P.: Events following the infection of enucleate cells with measles virus. J. gen. Virol. *32:* 163–175 (1976).

129 Hodes, D.S.: Temperature sensitivity of subacute sclerosing panencephalitis virus and its ability to establish persistent infection, Proc. Soc. exp. Biol. Med. *161:* 407–411 (1979).

130 Hamilton, R.; Barbosa, L.; Dubois, M.: Subacute sclerosing panencephalitis measles virus: study of biological markers. J. Virol. *12:* 632–642 (1973).

131 Albrecht, P.; Schumacher, H.P.: Neurotropic properties of measles virus in hamsters and mice. J. infect. Dis. *124:* 86–93 (1971).

132 Janda, Z.; Norrby, E.; Marusyk, H.: Neurotropism of measles virus variants in hamsters. J. infect. Dis. *124:* 553–564 (1972).

133 Wear, D.J.; Rapp, F.: Encephalitis in newborn hamsters after intracerebral injection of attenuated human measles virus. Nature, Lond. *227:* 1347–1348 (1970).

134 Woyciechowska, J.; Breschkin, A.M.; Rapp, F.: Measles virus meningoencephalitis. Immunofluorescence study of brains infected with virus mutants. Lab. Invest. *36:* 233–236 (1977).

135 Imagawa, D.T.; Adams, J.M.: Propagation of measles virus in suckling mice. Proc. Soc. exp. Biol. Med. *98:* 567–569 (1958).

136 Wear, D.J.; Rabin, E.R.; Richardson, L.S.; Rapp, F.: Virus replication und ultrastructural changes after induction of encephalitis in mice by measles virus. Exp. molec. Path. *9:* 405–417 (1968).

137 Sergiev, P.G.; Ryazantseva, N.E.; Shroit, I.G.: The dynamics of pathological processes in experimental measles in monkeys. Acta virol., Prague *4:* 265–273 (1960).

138 Byington, D.P.; Johnson, K.P.: Experimental subacute sclerosing panencephalitis in the hamster: correlation of age with chronic inclusion-cell encephalitis. J. infect. Dis. *126:* 18–26 (1972).

139 Byington, D.P.; Johnson, K.P.: Subacute sclerosing panencephalitis (SSPE) agent in hamsters. II. The neuropathology of acute and chronic infections. Exp. molec. Path. *18:* 345–356 (1973).

140 Byington, D.P.; Castro, A.E.; Burnstein, T.: Adaptation to hamsters of neurotrophic measles virus from subacute sclerosing panencephalitis. Nature, Lond. *225:* 554–555 (1970).

141 Griffin, D.E.; Mullinix, J.; Narayan, O.; Johnson, R.T.: Age dependence of viral expression: comparative pathogenesis of two rodent-adapted strains of measles virus in mice. Infect. Immunity *9:* 690–695 (1974).

142 Herndon, R.M.; Rena-Descalzi, L.; Griffin, D.E.; Coyle, P.K.: Age dependence of viral expression. Electron microscopic and immunoperoxidase studies of measles virus replication in mice. Lab. Invest. *33:* 544–553 (1975).

143 Waksman, B.H.; Burnstein, T.; Adams, R.D.: Histologic study of the encephalomyelitis produced in hamsters by a neurotropic strain of measles. J. Neuropath. exp. Neurol. *21:* 25–49 (1962).

144 Johnson, K.P.; Swoveland, P.: Measles antigen distribution in brains of chronically infected hamsters: an immunoperoxidase study of experimental subacute sclerosing panencephalitis. Lab. Invest. *37:* 459–465 (1977).

145 Albrecht, P.; Shabo, A.L.; Burns, G.R.; Tauraso, N.M.: Experimental measles encephalitis in normal and cyclophosphamide-treated rhesus monkeys. J. infect. Dis. *126:* 154–161 (1972).

146 Johnson, K.P.; Byington, D.P.: Subacute sclerosing panencephalitis (SSPE) agent in hamsters. I. Acute giant cell encephalitis in newborn animals. Exp. molec. Path. *15:* 373–379 (1971).

147 Johnson, K.P.; Norrby, E.: Subacute sclerosing panencephalitis (SSPE) agent in hamsters. III. Induction of defective measles infection in hamster brain. Exp. molec. Path. *21:* 166–178 (1974).

148 Lehrich, J.R.; Katz, M.; Rorke, M.B.; Barbanti-Brodano, G.; Koprowski, H.: Subacute sclerosing panencephalitis in hamsters produced by viral agents isolated from human brain cell. Archs. Neurol., Chicago *23:* 97–102 (1970).

149 Katz, M.; Rorke, L.B.; Masland, W.S.; Koprowski, H.; Tucker, S.H.: Transmission of an encephalitogenic agent from brain of patients with subacute sclerosing panencephalitis to ferrets. Preliminary report. New Engl. J. Med. *279:* 793–798 (1968).

150 Katz, M.; Rorke, L.B.; Masland, W.S.; Brodano, G.B.; Koprowski, H.: Subacute sclerosing panencephalitis: isolation of a virus encephalitogenic for ferrets. J. infect. Dis. *121:* 188–195 (1970).

151 Byington, D.P.; Burnstein, T.: Measles encephalitis produced in suckling rats. Exp. molec. Path. *19:* 36–43 (1973).

152 Katz, M.; Kackell, Y.; Muller, D.; ter Meulen, V.; Koprowski, H.: Immunohistological, microscopical, and neurohistochemical studies on encephalitides. VII. Subacute sclerosing panencephalitis. Susceptibility of dogs to the virus and characterization of host response. Acta neuropath. *25:* 81–88 (1973).

153 Greenham, L.W.; Peacock, D.B.; Hill, T.J.; Brownell, B.; Schutt, W.H.: The isolation of SSPE measles virus in newborn mice. Arch. ges. Virusforsch. *44:* 109–120 (1974).

154 Byington, D.P.; Johnson, K.P.: Subacute sclerosing panencephalitis virus in immunosuppressed adult hamsters. Lab. Invest. *32:* 91–97 (1975).

155 Haspel, M.V.; Duff, R.; Rapp, F.: Experimental measles encephalitis: a genetic analysis. Infect. Immunity *12:* 785–790 (1975).

156 Haspel, M.V.; Duff, R.; Rapp, F.: Isolation and preliminary characterization of temperature-sensitive mutants of measles virus. J. Virol. *15:* 1000–1009 (1975).

157 Burkholtz, C.M.; Kiley, M.P.; Payne, F.E.: Isolation and characterization of temperature-sensitive mutants of measles virus. J. Virol. *16:* 192–202 (1975).

158 Norrby, E.; Swoveland, P.; Kristensson, K.; Johnson, K.P.: Further studies on subacute encephalitis and hydrocephalus in hamsters caused by measles virus from persistently infected cell cultures. J. med. Virol. *5:* 109–116 (1980).

159 Kiley, M.P.; Payne, F.E.: Evidence of precursors of defective measles virus. Med. Microbiol. Immunol. *160:* 91–97 (1974).

160 Kiley, M.P.; Gray, R.H.; Payne, F.E.: Replication of measles virus: distinct species of short nucleocapsids in cytoplasmic extracts of infected cells. J. Virol. *13:* 721–728 (1974).

161 Cernescu, C.; Sorodoc, Y.: Subacute sclerosing panencephalitis and defective interfering measles virus particles. Revue roum. med. Virol. *31:* 3–8 (1980).

162 Rima, B.K.; Davidson, W.B.; Martin, S.J. The role of defective interfering particles in persistent infection of Vero cells by measles virus. J. gen. Virol. *35:* 89–97 (1977).

163 Huang, A.S.; Baltimore, D.: Defective interfering animal viruses; in Fraenkel-Conrat, Wagner, Comprehensive virology, vol. 10, pp. 73–116 (Plenum Press, New York 1977).

164 Ahmed, R.; Chakraborty, P.R.; Fields, B.N.: Genetic variation during lytic reovirus infection: high-passage stocks of wild-type reovirus contain temperature-sensitive mutants. J. Virol. *34:* 285–287 (1980).

165 Rustigian, R.: A carrier state in HeLa cells with measles virus (Edmonston strain) apparently associated with noninfectious virus. Virology *16:* 101–104 (1962).

166 Rustigian, R.: Persistent infection of cells in culture by measles virus. I. Development and characteristics of HeLa sublines persistently infected with complete virus. J. Bact. *92:* 1794–1804 (1966).

167 Rustigian, R.: Persistent infection of cells in culture by measles virus. II. Effect of measles antibody on persistently infected HeLa sublines and recovery of a HeLa clonal line persistently infected with incomplete virus. J. Bact. *92:* 1805–1811 (1966).

168 Barry, D.W.; Sullivan, J.L.; Lucas, S.J.; Dunlap, R.C.; Albrecht, P.: Acute and chronic infection of human lymphoblastoid cell lines with measles virus. J. Immun. *116:* 89–98 (1976).

169 Gibson, P.E.; Bell, T.M.: Persistent infection of measles virus in mouse brain cell cultures infected in vivo. Arch. ges. Virusforsch. *37:* 45–53 (1972).

170 Gould, E.A.; Linton, P.E.: The production of temperature-sensitive persistent measles virus infection. J. gen. Virol. *28:* 21–28 (1975).

171 Haspel, M.V.; Knight, P.R.; Duff, R.G.; Rapp, F.: Activation of a latent measles virus infection in hamster cells. J. Virol. *12:* 690–695 (1973).

172 MacIntyre, E.H.; Armstrong, J.A.: Fine structural changes in human astrocyte carrier lines for measles virus. Nature, Lond. *263:* 232–234 (1976).

173 Minagawa, T.: Studies on the persistent infections with measles virus in HeLa cells. I. Clonal analysis of cells in carrier cultures. Jap. J. Microbiol. *15:* 325–331 (1971).

174 Minagawa, T.; Sakuma, T.; Kuwajima, S.; Yamamoto, T.K.; Iida, H.: Characterization of measles viruses in establishment of persistent infections in human lymphoid cell line. J. gen. Virol. *33:* 361–379 (1976).

175 Norrby, E.: A carrier cell line of measles virus in LU 106 cells. Arch. ges. Virusforsch. *20:* 215–224 (1967).

176 Knight, P.; Duff, R.; Rapp, F.: Latency of human measles virus in hamster cells. J. Virol. *10:* 995–1001 (1972).

177 Knight, P.; Duff, R.; Glaser, R.; Rapp, F.: Characteristics of the release of measles virus from latently infected cells after co-cultivation with BSC-1 cells. Intervirology *2:* 287–298 (1973).

178 Menna, J.H.; Collins, A.R.; Flanagan, T.D.: Characterization of an in vitro persistent-state measles virus infection: establishment and virological characterization of the BGM/MV cell line. Infect. Immunity *11:* 152–158 (1975).

179 Wild, T.F.; Durge, R.: Establishment and characterization of a subacute sclerosing panencephalitis (measles) virus persistent infection in BGM cells. J. gen. Virol. *39:* 113–124 (1978).

180 Kohno, S.; Kohase, M.; Suganuma, M.: Growth of measles virus in a mouse-derived established cell line, L cells. Jap. J. med. Sci. Biol. *21:* 301–311 (1968).

181 Rentier, B.; Dubois-Dalcq, M.; Claysmith, A.P.; Bellini, W.J.: Chronic measles virus infection of nerve cells in vitro; in Bishop, Compans, Replication of negative strand viruses (Elsevier/North Holland, Amsterdam, in press, 1981).

182 Payne, F.E.; Baublis, J.V.: Decreased reactivity of SSPE strains of measles virus with antibody. J. infect. Dis. *127:* 505–511 (1973).

183 Horta-Barbosa, L.; Fuccillo, D.A.; Hamilton, R.; Traub, R.; Ley, A.; Sever, J.L.: Some characteristics of SSPE measles virus. Proc. Soc. exp. Biol. Med. *134:* 17–21 (1970).

184 Payne, F.E.; Baublis, J.V.: Measles virus and subacute sclerosing panencephalitis. Perspect. Virol., vol. 7, pp. 179–185 (Academic Press, New York 1971).

185 Machamer, C.E.; Hayes, E.C.; Gollobin, S.D.; Westfall, L.K.; Zweerink, H.J.: Antibodies against the measles matrix polypeptide after clinical infection and vaccination. Infect. Immunity *27:* 817–825 (1980).

186 Trudgett, A.; Bellini, W.J.; Mingioli, E.S.; McFarlin, D.E.: Antibodies to the structural polypeptides of measles virus following acute infection and in SSPE. Clin. exp. Immunol. *39:* 652–656 (1980).

187 Hayes, E.C.; Gollobin, S.D.; Machamer, C.E.; Westfall, L.K.; Zweerink, H.J.: Measles-specific antibodies in sera and cerebrospinal fluids of patients with multiple sclerosis. Infect. Immunity *27:* 1033–1037 (1980).

188 Fujinami, R.S.; Oldstone, M.B.A.: Alterations in expression of measles virus polypeptides by antibody: molecular events in antibody-induced antigenic modulation. J. Immun. *125:* 78–85 (1980).

189 Joseph, B.S.; Oldstone, M.B.A.: Immunologic injury in measles virus infection. J. exp. Med. *142:* 864–876 (1975).

Dr. Steven L. Wechsler, Department of Molecular Virology, Christ Hospital Institute of Medical Research, Cincinnati, OH 45219 (USA)

Prog. med. Virol., vol. 28, pp. 96–113 (Karger, Basel 1982)

Korean Hemorrhagic Fever[1]

With 1 color plate

Ho Wang Lee

World Health Organization Collaborating Centre for Research on Korean Haemorrhagic Fever, The Institute for Viral Diseases, Korea University, Seoul, Korea

Contents

[1] Aided by grant No. DAMD17-80-G-9468 from the US Army Medical Research and Development Command, Frederick, MD 21701, USA.

I. Introduction

An epidemic of hemorrhagic fever with an associated renal syndrome attracted great attention during the Korean War when several thousand hospitalized cases were reported among United Nations troops [8, 47]. Since that time, this disease has become known as Korean hemorrhagic fever [KHF]. South Korea-wide, approximately 300–900 persons are hospitalized with KHF every year [28, 29, 34].

Hemorrhagic fevers with a syndrome very similar to KHF have been reported throughout Euro-Asia: *(i)* as hemorrhagic nephroso-nephritis or hemorrhagic fever with renal syndrome in the Soviet Union [5, 6, 48, 49], with several thousand cases reported annually since 1913; *(ii)* as Songo fever or epidemic hemorrhagic fever (EHF) in China [16, 19, 23], with more than 20,000 cases reported annually since 1931; *(iii)* as nephropathia epidemica (NE) in Scandinavia [25, 45], with several hundred cases reported annually since 1934; *(iv)* as epidemic nephritis or epidemic hemorrhagic fever in Eastern Europe [11, 12, 55], since 1934; and *(v)* as epidemic hemorrhagic fever in Japan [52], since 1960.

In addition to these, several thousand cases of a War nephritis clinically similar to NE were reported among British soldiers stationed in Flanders during World War I [1, 2, 4, 17]; about 14,000 cases of War nephritis were also described among Northern Armies of the Central Region in the American Civil War [4, 17].

In 1976, *Lee and Lee* [26] discovered a specific antigen of KHF from the striped field mouse, *Apodemus agrarius*. In 1978, *Lee* et al. [28] demonstrated that this antigen is the etiologic agent of KHF and perfected a serologic test for diagnosis of the disease. These results came after five decades of virtually continuous but heretofore unrewarding research by several groups of investigators.

Very recently, close etiological relations have been proved by serologic means between KHF in Korea and hemorrhagic nephroso-nephritis in the USSR [28], between KHF and NE in Scandinavia [9, 32, 50, 51], between KHF and EHF in Japan [31] and between KHF and EHF in China [42–44]. Recent sero-epidemiologic surveys show that agents closely related and antigenically similar to Hantaan virus, the etiologic agent of KHF, are widely distributed throughout much of the world, to a much greater geographic extent than previously recognized (fig. 1). The existence of a KHF-related agent in Flanders and in the USA will be an interesting topic for future study.

The present paper is a review of recent progress in Korean hemorrhagic fever research.

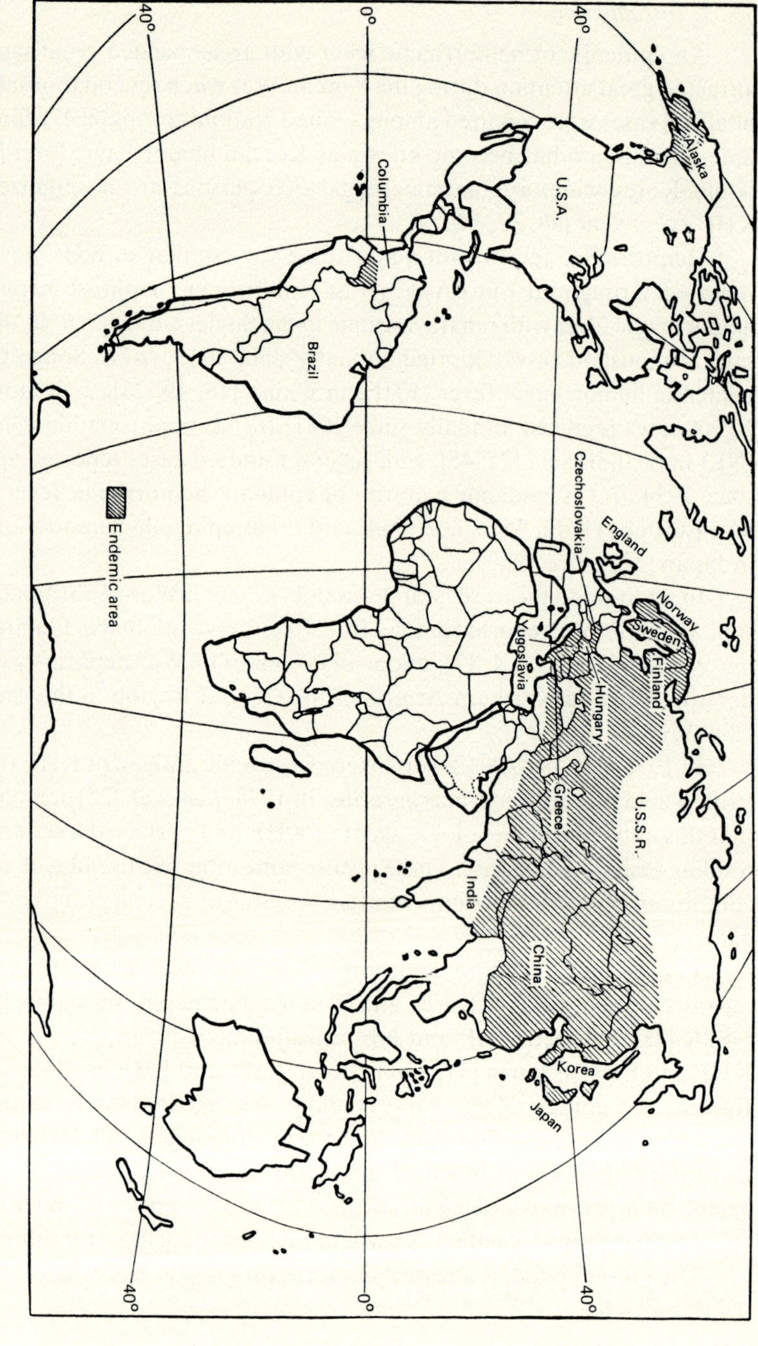

Fig. 1. World distribution of Hantaan virus, the etiologic agent of KHF, based upon demonstration of antibody in man against the virus.

II. Etiology

A. Background

In the early 1940s, scientists in the Soviet Union reproduced the disease by inoculation of sera and urine from patients into volunteers [49]. In this same decade, Japanese scientists working in China reported that they were able to reproduce the disease in monkeys by inoculation of sera from patients and suspensions of mites from *Apodemus* mice [19, 23]. Filtered sera from patients also produced clinical symptoms, therefore the disease has for many years been suspected of being caused by a virus.

Concerning the preliminary Japanese studies [19, 23], it has been speculated that humans rather than monkeys may have been used as experimental hosts [17, 18, 53]. Later efforts have all failed to reproduce the disease in monkeys.

Gajdusek reviewed the available world literature on virus hemorrhagic fevers in 1962 [11], and he postulated that hemorrhagic fevers with renal syndrome occurring across the northern portion of Euro-Asia, the world's largest continent with more than two billion people, may be caused by a single etiologic agent.

As *Traub* [54] described, rarely has an infectious disease been investigated as intensively as was KHF in the 1951–1954 period by workers on both the Russian and American sides. Many attempts, then and since, were made to isolate the etiologic agent of KHF and clinically similar diseases in both tissue culture cells and in experimental animals including monkeys. A Russian report of cultivation of an agent in cell cultures from blood specimens of patients with hemorrhagic nephroso-nephritis [13] has not been confirmed; other Russian claims as to isolation of a virus from a patient were not deemed definitive [6], and an agent that produced cytopathic effect (CPE) in pig embryo kidney cells was lost [5].

B. Isolation of Hantaan Virus, the Etiologic Agent of KHF

In 1976, *Lee and Lee* [26] demonstrated an antigen in the lungs of wild *Apodemus agrarius coreae* captured in rural endemic areas of KHF that gave specific immunofluorescent [IF] reactions with convalescent sera from KHF patients. In 1978, *Lee* et al. [28] reported that this antigen is the etiologic agent of KHF, discovered the animal reservoir host and perfected a serologic test for diagnosis of the disease. Viral strains also have since been recovered from the blood of KHF patients.

The agent was named Hantaan virus after the Hantaan river which runs near Songnaeri where *Lee* et al. [28] first isolated the virus from *Apodemus* mice, and this name has been so registered in the Catalogue of Arthropod-borne Viruses and Selected Vertebrate Viruses of the World [15].

Very recently, *Brummer-Korvenkontio* et al. [3] reported recovery of an agent similar to KHF from the pulmonary tissues of *Chlethrionomys glareolus* and which appears to be etiologically linked to a clinical syndrome termed nephropathia epidemica in Finland. The etiologic relationship between Hantaan virus and other similar hemorrhagic fevers of unknown etiology that occur in Eastern Europe remains to be determined.

C. Growth of Virus and Animal Susceptibility

Hantaan virus grows well both in vivo and in vitro. *Lee* et al. [40] conducted studies to define the natural host range of Hantaan virus and to identify colonized rodents susceptible to this infection. Out of 9 species tested among field rodents in the rural endemic areas of KHF, only the species *Apodemus agrarius* showed the presence of antigen in their tissues. When 12 species of colonized laboratory rodents were inoculated with virus, only two, *Apodemus agrarius ninpoensis* and *Calomys callosus,* developed detectable virus antigen in their tissues [40].

When the virus is inoculated into *Apodemus* mice, the virus begins to appear in the lungs after 10 days. Thereafter it can be identified in kidneys, liver and submaxillary glands. Maximal amounts of virus can be detected in the lungs on or about the 20th day, after which quantities decline gradually; after 150 days it is still possible to detect the virus.

Neither wild-caught nor laboratory-inoculated *Apodemus agrarius* ever display overt signs of disease [28, 41]. Histological examination of infected rodent tissues does not show striking inflammatory lesions, but blood analysis of infected rodents shows a marked leukocytosis [33].

French et al. [10] succeeded recently in propagating the virus in an in vitro system for the first time in a human cell line designated A-549 and described as type II, alveolar epithelial cells derived from carcinoma of the lung. The virus after six passages in the cells produced specific fluorescence in the cytoplasm of the cells on day 7 postinoculation. This finding makes it possible for the first time to study KHF-related antigen in infected cells utilizing IF antibody techniques.

Table I. Physicochemical properties of Hantaan virus, the etiologic agent of Korean hemorrhagic fever

Description	Properties
Size	approx. 75 nm
Morphology	spherical?
Filtration	
100 nm	passed
50 nm	not passed
BUDR	resistant
Deoxycholate (0.1%)	inactivated
Ether	inactivated
Chloroform	inactivated
Acetone	inactivated
Benzene	inactivated
Ultraviolet light	inactivated
pH 5	inactivated
pH 9	stable
Heat 56 °C, 30 min	inactivated
0.5 *N* HCl	inactivated
0.05 *N* NaOH	inactivated
0.5% iodine	inactivated
70% ethanol	inactivated

D. Properties of Hantaan Virus

Hantaan virus is a small RNA virus [15], and study of the morphology of this virus is in progress at several laboratories. The virus passes through a 100-nm millipore filter but not a 50-nm filter and thus has a diameter of approximately 75 nm. The virus is sensitive to deoxycholate, ether and chloroform.

Infectivity of the virus to *Apodemus* mice is stable at pH 7.0–9.0 but is inactivated completely at pH 5.0. The virus is relatively stable at 4–20 °C but is inactivated rapidly at 37 °C. The seed virus in a diluent containing 1% bovine albumin in balanced salt solution (BSS) can be stored at −60 °C for 4 years at least [15]. Some of the physicochemical characteristics of the virus are summarized in table I.

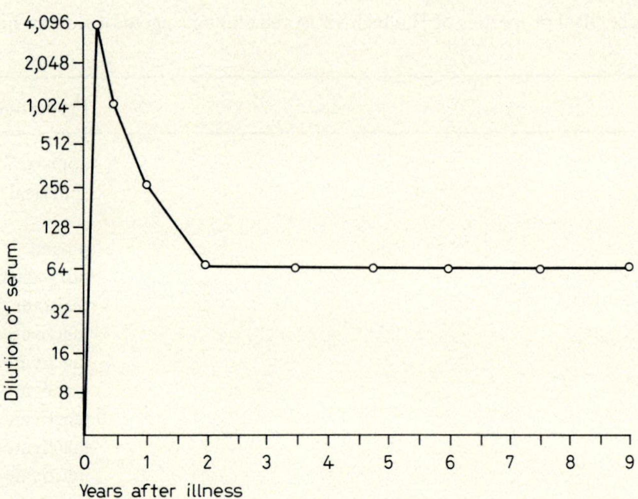

Fig. 2. Persistence of indirect fluorescent antibodies to Hantaan virus in the sera of a patient with KHF.

III. Immunology

Lee et al. [27, 37] reported results of a quantitative analysis of complement, immunoglobulins and serum proteins of KHF patients during the course of illness. The immunoglobulin which appears first in the patient's serum is IgM. IgM increases abruptly in the early phase, reaches a peak during days 7–11 after onset of fever, then falls off rapidly to initial levels. This is followed by the appearance and rising titers of IgG and IgA. IgG reaches peak levels during days 10–14 after onset of fever. At this same time the immunofluorescent (IF) and neutralizing (N) antibodies against Hantaan virus are produced. Some of the patients show increased levels of IgE during days 11–16, thereafter falling off slowly. The serum levels of C3 and C5 are depressed during days 3–7, thereafter rising back to normal levels. Increased levels of ceruloplasmin and alpha-antitrypsin are observed during the early febrile stage of the illness [37].

The IF antibodies begin to appear right after onset of fever, and the highest titers are observed at 2–3 weeks, followed by a slow decline. In most but not all cases the antibodies appeared by the 7th day. The IF and N antibodies to Hantaan virus were present in each of 13 sera obtained from patients with KHF 3–14 years [28, 36], and in two sera obtained from patients with Songo fever in China 34 years, after acute illness [42]. Persistence and levels of IF antibodies to the virus in the sera of a KHF patient after illness are shown in figure 2.

Occurrence of IF antibodies to Hantaan virus in the sera from Korean residents of endemic areas, near-endemic areas and non-endemic areas of KHF was 3.8 %, 2.7 %, and 1.0 %, respectively [28, 30]. All of the seropositive samples belonged in the age group of 20–50 years and no difference was observed by sex [28, 30, 34]. Specific antibodies have been identified in the sera not only from individuals showing severe clinical symptoms but also from those having mild and subclinical symptoms [36]. Immunity to KHF after recovery from the acute illness appears to be life-long, since antibodies to Hantaan virus have been shown to persist as long as 34 years and since there has been no case of reinfection yet in endemic areas.

Experimental animals, Swiss albino mice, rats (Wistar), guinea pigs, rabbits and monkeys produce IF and N antibodies after inoculation with a single dose of Hantaan virus intramuscularly [40]. Lack of any immunological relationship between Hantaan virus and antisera of arenaviruses, Lassa, Ebola and other known viruses was proved by the indirect fluorescent antibody (FA) test method utilizing infected *Apodemus* lungs and infected A-549 cells as substrate and by the complement fixation test method [15].

IV. Laboratory Diagnosis

It is impossible in the individual case with moderate to mild clinical symptoms to diagnose Hantaan virus infection on clinical grounds alone, and even in severe cases an accurate clinical diagnosis of this disease can be difficult. In our limited study, a correct clinical diagnosis was made in only 50% of seropositive KHF patients in Korea [36]. The IF technique using infected *Apodemus* lung sections and A-549 cells with Hantaan virus [10, 28] makes it possible to diagnose serologically patients with KHF, hemorrhagic nephroso-nephritis in the USSR, EHF in Japan and China, NE in Finland and benign nephropathy in Sweden. The procedure of choice for the definitive diagnosis of Hantaan virus infection is the demonstration of rising titers of IF antibodies to Hantaan virus in the sera from suspected patients.

A. Specific Serological Diagnosis

Specific serological diagnosis of KHF [28] is made by demonstrating a rise in titer of specific IF antibodies and N antibodies against Hantaan virus in sera collected twice during the course of illness at an interval of 1 week. IF and N antibodies appear during the 1st week of symptoms, reach a peak at the end of the 2nd week, and persist for as long as 34 years. Specific antibodies to the virus can be detected even in cases with mild and subclinical symptoms.

IF antibodies against Hantaan virus can be detected utilizing virus-infected *Apodemus* lung sections or virus-infected A-549 cells [10] as substrate. Specific fluorescence appears as discrete pinpoint granules distributed throughout the cytoplasm of the cells. The neutralization test in *Apodemus* mice is specific but it is expensive and time-consuming.

An indirect IF test for detection of specific antibodies for NE has been developed with lung sections of bank vole that had been trapped from the endemic areas of NE in Finland as antigen in 1980 [3].

B. Isolation of the Virus

Isolation of Hantaan virus is possible by inoculation of *Apodemus* mice or A-549 cells with blood and serum taken in the early stage of infection. *Apodemus* mice infected with the virus do not show overt signs of disease, but the viral antigen can be recognized by observing the pulmonary tissues by means of the indirect FA technique. To date, we have isolated 19 strains of Hantaan virus in *Apodemus* mice from 105 blood specimens from acute stage patients with KHF.

V. Epidemiology of Korean Hemorrhagic Fever

KHF was not described in Korea prior to 1951, although this by no means disproves the existence of the disease on the Korean peninsula before this time. The disease in fact may have existed in Asia for at least the last 1,000 years, inasmuch as there is a suggestive description of a hemorrhagic fever with an associated renal syndrome in a Chinese Medicine book (Whang-Jae-Nae-Kyung) which was written around the year 960 A.D.

During the Korean War KHF was an important military problem, having been responsible for several thousand cases among US troops, not to mention those in opposing forces. Outbreaks of KHF had been localized in the vicinity of the demilitarized zone (38th parallel) between South and North Korea up until 1970, but since 1971 the disease has invaded the southern areas of the Korean peninsula [21, 22, 29]. Currently, the numbers of afflicted civilian patients are increasing more rapidly than those of the military services, and the disease is now prevalent all over the Korean peninsula; new cases are being reported not only in rural areas but in urban areas as well [38, 52].

A. Epidemiologic Types

There are three epidemiologic types of KHF according to the outbreaks and the reservoir host of the disease.

(1) *Rural Type.* The reservoir for KHF in rural areas of Korea, China, and probably in the Eastern parts of the USSR as well, is *Apodemus* species [28, 30], while the reservoir for NE in Scandinavia is *C. glareolus* [3]. These rodents live only in the field but will invade houses during the snowing season. Cases of the rural type occur in the endemic rural areas, and annually there are two seasonal peaks when incidence of the disease is high.

(2) *Urban Type.* The reservoir is house rats. There were many cases of KHF reported in metropolitan areas of Seoul in individuals who have never been outside the city limits and who had histories of contact with house rats [38]; 130 cases of KHF were reported also among residents of urban areas of Osaka City in Japan [52]. Cases of the urban type occur throughout the year but tend to appear more in fall and winter seasons. We have evidence that house rats are indeed a reservoir of Hantaan virus by the demonstration of antibodies to KHF [in preparation].

(3) *Experimental Animal Type.* The reservoir is colonized rats [56]. There were several outbreaks of KHF among personnel of laboratory animal rooms in Korea and in Japan, and it was proved that the rat (Wistar) was the source of infection. These cases may occur at any time of the year, but a series of outbreaks occurred during the winter season when the air is dry in the animal room of our Institute.

B. Seasonal Incidence

KHF occurs throughout the year, but the incidence is greatest during two sharp seasonal peaks, a small one in the late spring and a larger peak in the late fall [8, 14, 29, 30] as shown in figure 3. Cases usually appear singly, but small outbreaks do develop, particularly when groups of susceptible persons are exposed to a contaminated focus, such as digging burrows of infected rodents.

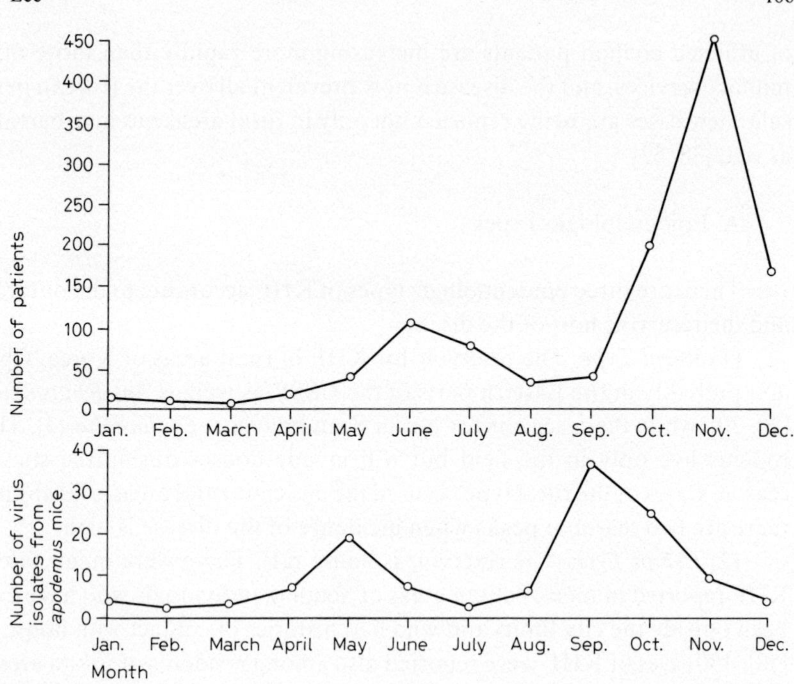

Fig. 3. Cumulative seasonal prevalence of KHF and of *Apodemus agrarius* infected with Hantaan virus in the endemic area of Kyunggi Province in Korea from 1975 to 1978.

The interaction between the seasonal copulatory activities of *Apodemus* mice and the seasonal planting and harvesting activities of the farmer produce the seasonal incidence pattern observed in the rural type of KHF. A large portion of *Apodemus* mice captured wild from endemic areas during the month of June and during the months of September–October are found to be infected [28, 30], as shown in figure 3. It is during these periods when *Apodemus* mice are reproductively active; the mice come outside from their burrows to copulate [39]. Furthermore, as gravid mice are known to excrete much larger quantities of excreta than non-gravid mice, much more infective excreta are deposited on above-ground sites by *Apodemus* mice during these periods.

During the late spring and the late fall, in connection with the usual activities of planting and harvesting of grain, the farmer spends much time in the fields which is where he is most likely to come in contact with infective excreta. The dry and dusty air of the late spring and late fall, corresponding to the dry seasons in Korea, is also a contributing factor, inasmuch as sun-dried excreta may become airborne and then as minute particulate matter be aspirated by humans. Finally, in the fall season after the harvest, *Apodemus* mice are known to come into the villages in search of food, thus during this time coming into closer than usual contact with humans.

C. Age and Sex

The disease appears to affect most frequently the age group of 20–50 years; cases under age 10 years are rare. Although KHF occurs in both sexes, the incidence figures accumulated thus far show a significantly higher incidence in males [21, 22]. The victims are primarily farmers and soldiers stationed in the field.

D. Animal Reservoir Host

The reservoir of KHF in the rural endemic areas in Korea is *Apodemus agrarius coreae* [28]. There are 9 species of field rodents [20] in the endemic areas, but only *Apodemus* species harbor the Hantaan virus [8, 28, 30, 40]. Infected rodents excrete large amounts of virus in saliva, in urine and in feces for a long period of time [41]. The reservoir of NE in Finland and west of the Ural mountains is *C. glareolus* [3, 5, 57]. The reservoir of KHF in the urban areas of Korea is house rats. Colonized experimental rats are a dangerous reservoir of KHF which have been responsible for several outbreaks of KHF among personnel of laboratory animal rooms at the Universities of Tohoku [56], Niigata, Wakayama and Nagoya in Japan [in preparation].

E. Virus Transmission in the Reservoir Animal

It has been hypothesized that KHF can be transmitted by ecto-parasites harbored by various field rats [19, 23, 24, 48, 53, 54]. The only positive report supportive of this view so far published was that by *Kasahara* et al. [19] who were able to induce the disease in monkeys by inoculation of a saline suspension of 203 mites, *Laelaps jettmari Vitzthum,* obtained from 40 *Apodemus agrarius.* However, to date this finding has yet to be confirmed by independent investigators though many attempts have already been made.

Lee et al. [41] demonstrated for the first time that large quantities of virus are excreted in the saliva, in the urine and in the feces from infected *Apodemus agrarius.*

The infected *Apodemus* mice excrete the virus in saliva and the feces for a period of 1 month and in the urine for 12 months at least. Horizontal transmission of the virus between *Apodemus* mice was demonstrated as non-infected *Apodemus* mice caged together with infected *Apodemus* for several days acquired infection beginning 10 days after inoculation of the virus into the first animal. Results were not different when ectoparasitized and clean animals were used in these experiments. The main route of infection in *Apodemus* mice is through the respiratory tract, and infection can be transmitted via the saliva, urine or feces of infected *Apodemus* mice. Transmission of the virus among house rats and experimental animals remains to be studied.

VI. Clinical Course

The recent development of a definitive serological diagnosis for KHF has facilitated the recognition that the clinical disease produced by the Hantaan virus is one of diverse and protean manifestations. Before serodiagnosis, KHF was diagnosed on clinical grounds alone, its major manifestations being fever, prostration, vomiting, proteinuria, hemorrhagic phenomena, shock and renal failure. The incubation period is generally 2–3 weeks, but may vary from 4 to 42 days. Mild and moderate forms of the disease, without hemorrhagic phenomena or proteinuria, occur frequently, and it is in these cases especially that serological diagnosis is recommended.

Approximately 30–40% of patients show a mild clinical course, about 50–60% exhibit a moderate course, while 20–30% have severe forms of the disease. The severe forms can be divided into five phases according to characteristic clinical, laboratory and pathophysiologic features, though in some patients the disease does not conform to the arbitrary categories now to be set forth [8, 35, 46, 47].

A. Febrile Phase

This phase usually lasts 3–7 days with a sudden onset of high fever, chills, general malaise, weakness and generalized myalgia, usually followed by severe anorexia, dizziness, headache and eyeball pain. The extensive retroperitoneal or peritoneal edema, resulting from extravasation of plasma,

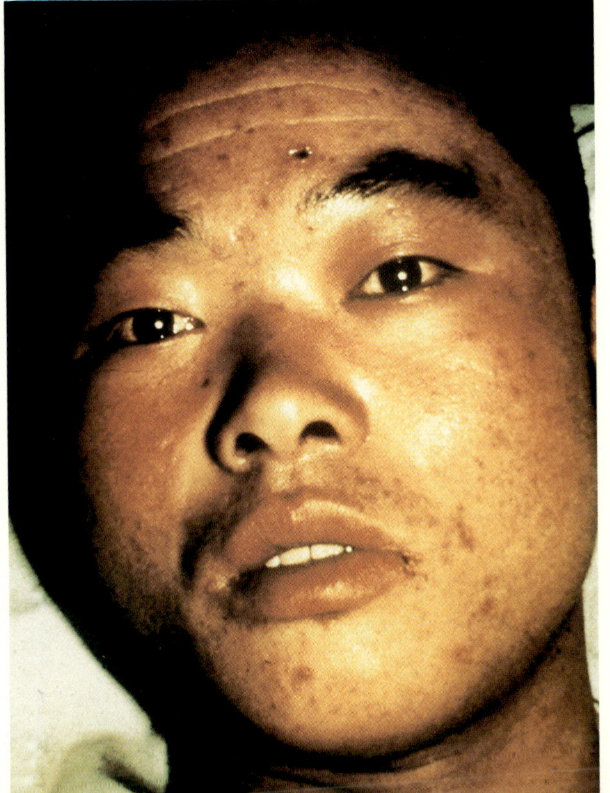

1

For legends see reverse side

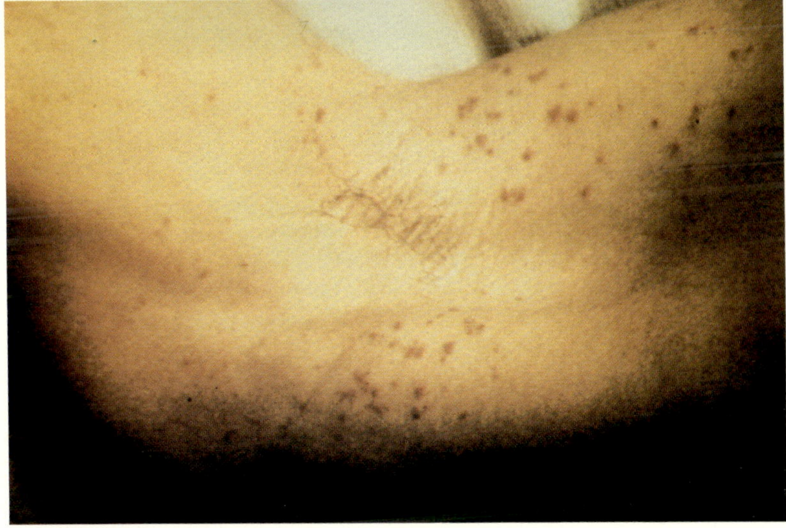

5

Fig. 4. Classic picture of KHF in a 21-year-old man 6 days after the onset of fever. Note the flushed face and conjunctival hemorrhage.

Fig. 5. Petechiae in the axillary region of the patient shown in figure 4.

accounts for the severe abdominal and back pain and the tenderness in the renal area. Thereafter, the characteristic findings of KHF, namely, flushing over the face, neck and anterior chest with injection of the eyes, palate and pharynx, occur (fig. 4). Toward the end of this phase fine petechiae are observed in the axillary folds (fig. 5), and on the face, neck, soft palate and anterior chest wall, together with conjunctival hemorrhage. During the early febrile phase, the urine may contain a small amount of albumin, which then increases abruptly in the late febrile phase reaching massive proteinuria in the majority of cases.

B. Hypotensive Phase

This phase develops abruptly and lasts several hours to 2 days. The classic picture of shock, including tachycardia, narrowed pulse pressure, hypotension, cold and clammy skin, dulled sensorium and even confusion may be noted.

In severe forms of KHF, one-third of fatal cases are associated with irreversible shock in this stage. Laboratory data show marked proteinuria and mild hematuria with urine specific gravity beginning to fall. The hematocrit rises, a leukemoid reaction is found and the number of platelets decreases. Capillary hemorrhages are most prominent at this time. Oliguria may appear during the late shock phase, following which the blood urea and creatinine begin to rise.

C. Oliguric Phase

This phase lasts 3–7 days. The blood pressure begins to normalize but many patients (up to 60%) become hypertensive due to a relative hypervolemic state. Patients may show very severe nausea and vomiting associated with persistant oliguria, but facial flushing and petechiae begin to disappear. Bleeding tendencies increase markedly, such that about one-third of patients show various combinations of epistaxis, conjunctival hemorrhage, cerebral hemorrhage, gastrointestinal bleeding and extensive purpura. Hyperkalemia, hyponatremia and hypocalcemia may also be present, but metabolic acidosis is rarely seen. Central nervous system symptoms and pulmonary edema may occur in the severe forms of the disease, and about 50% of all case fatalities occur during this oliguric phase.

D. Diuretic Phase

With the onset of the diuretic phase, which may last for days or weeks, clinical recovery is initiated. Diuresis may be delayed due to dehydration, electrolyte imbalance and infection in some patients. Diuresis of 3–6 liters daily is the rule. The daily urine output and the duration of the diuretic phase are influenced greatly by the severity of the disease.

E. Convalescent Phase

The convalescent phase requires 2–3 months; mild anemia may persist but recovery is the rule. This phase is characterized by a progressive recovery of glomerular filtration rate, renal blood flow and urine-concentrating ability, up to 70% of normal status within 6 months of illness.

References

1 Abercrombie, R.G.: Observations on the acute phase of five hundred cases of War nephritis. Jl R. Army med. Cps *27:* 10–157 (1916).

2 Bradford, J.R.: Nephritis in the British troops in Flanders. Q.Jl Med. *9:* 445–459 (1916).

3 Brummer-Korvenkontio, M.; Vaheri, A.; von Bonsdorff, C.H.; Vuorimies, J.; Manni, T.; Penttinen, K.; Oker-Blom, N.; Laehdvirta, J.: Nephropathia epidemica: detection of antigen in bank voles and serologic diagnosis of human infection. J. infect. Dis. *141:* 131–134 (1980).

4 Brown, W.L.: Trench nephritis. Lancet *i:* 391–395 (1916).

5 Casals, J.; Hoogstraal, H.; Johnson, K.M.; Shelokov, A.; Wiebenga, N.H.; Work, T.: A current appraisal of hemorrhagic fevers in the U.S.S.R. Am. J. trop. Med. Hyg. *15:* 751–764 (1966).

6 Casals, J.; Henderson, B.; Hoogstraal, H.; Johnson, K.M.; Shelokov, A.: A review of Soviet viral hemorrhagic fevers, 1969. J. infect. Dis. *122:* 437–453 (1970).

7 Dodge, H.J.; Griffin, H.E.; Gauld, R.L.; Kim, Y.S.: Epidemic hemorrhagic fever in a Korean farm population. Am. J. Hyg. *63:* 38–51 (1956).

8 Earle, D.P. (ed.): Symposium on epidemic hemorrhagic fever. Am. J. Med. *16:* 617–709 (1954).

9 Friman, G.; French, G.R.; Hambraeus, L., and Beisel, W.R.: Scandinavian epidemic nephropathy and Korean hemorrhagic fever. Lancet *ii:* 100 (1980).

10 French, G.R.; Foulke, R.S.; Brand, O.A.; Eddy, G.A.; Lee, H.W.; Lee, P.W.: Propagation of etiologic agent of Korean hemorrhagic fever in a cultured continuous cell line of human origin. Science *211:* 1046–1048 (1981).

11 Gajdusek, D.C.: Virus hemorrhagic fever. J. Pediat. *60:* 851–857 (1962).

12 Gaon, J.; Karlovac, M.; Gresikova, M.; Hlaca, D.; Rukavania, J.; Kuezevic, V.; Saratlic-Savic, D.; Vampotic, A.: Epidemiological features of hemorrhagic fever. Folia med. Fac. Med. Univ. Saraev. *3:* 24–41 (1968).

13 Gavrilyuk, B.; Smorodintsev, A.A.: Isolation of hemorrhagic nephroso-nephritis virus in cell cultures. Arch. Virusforsch. *34:* 171–178 (1971).

14 Gauld, R.L.; Graig, J.P.: Epidemiological pattern of localized outbreaks of epidemic hemorrhagic fever. Am. J. Hyg. *59:* 32–38 (1954).

15 Information Exchange Subcommittee, American Committee on Arthropod-Borne Viruses (Karabatsos, ed.) Ft. Collins, Colo.

16 Ishii, S.; Ando, K.; Watanabe, N.; Murakami, R.; Nagayama, T.; Ishikawa, T.: Studies on Songo fever. Jap. Army med. J. *355:* 1755–1758 (1942).

17 Jellison, W.L.: Korean hemorrhagic fever and related diseases: a critical review and a hypothesis, pp. 1–79 (Mountain Press Pub. Co., Missoula, Mont., 1971).

18 Johnson, K.M.: Nephropathia epidemica and Korean hemorrhagic fever: the veil lifted? J. infect. Dis. *141:* 135–136 (1980).

19 Kasahara, S.; Kitano, M.; Kikuchi, H.; Sakyuama, M.; Kanazawa, K.; Nezu, N.; Yoshimura, M.; Kudo, T.: Studies on pathogen of epidemic hemorrhagic fever. J. Jap. Pathol. *34:* 3–5 (1944).

20 Kim, I.S.; Lim, Y.W.; Lee, H.W.: Studies of field rodents collected in the hyperendemic areas of Korean hemorrhagic fever. J. Korea med. Univ. *11:* 263–291 (1974)

21 Kim, K.H.; Antal, G.M.; Shin, H.K.: Epidemiological studies on hemorrhagic nephroso-nephritis (Korean hemorrhagic fever) in the Republic of Korea. Korean J. Virol. *4:* 29–39 (1974).

22 Kim, S.W.; Chun, J.H.: Changing patterns of epidemiology and death causes in epidemic hemorrhagic fever in Korea. Korean J. infect. Dis. *8:* 58–74 (1974).

23 Kitano, M.: A study of epidemic hemorrhagic fever. J. Manchou Med. *40;* 191–209 (1944).

24 Kwon, C.S.; Lim, Y.W.; Lee, H.W.: Studies on ectoparasites of field rodents collected in the endemic areas of Korean hemorrhagic fever. J. Korea med. Univ. *10:* 817–827 (1973).

25 Laehdvirta, J.: Nephropathia epidemica in Finland. Ann. clin. Res. *3:* 12–17 (1971).

26 Lee, H.W.; Lee, P.W.: Korean hemorrhagic fever. I. Demonstration of causative antigen and antibodies. Korean J. intern. Med. *19:* 371–384 (1976).

27 Lee, H.W.; Seong, I.W.; Kwon, B.H.: Qualitative analysis of immunoglobulins of Korean hemorrhagic fever patients. Korean J. Virol. *6:* 15–19 (1976).

28 Lee, H.W.; Lee, P.W.; Johnson, K.M.: Isolation of the etiologic agent of Korean hemorrhagic fever. J. infect. Dis. *137:* 298–308 (1978).

29 Lee, H.W.: Korean hemorrhagic fever; in Pattyn, Ebola virus hemorrhagic fever, pp. 331–343 (Elsevier/North Holland, Amsterdam 1978).

30 Lee, H.W.: Ecology of the etiologic agent of Korean hemorrhagic fever and its distribution in the world. Ann. Korean Acad. Scien. *17:* 177–209 (1978).

31 Lee, H.W.; Lee, P.W.; Tamura, M.; Tamura, T.; Okuno, Y.: Etiological relation between Korean hemorrhagic fever and epidemic hemorrhagic fever in Japan. Biken's J. *22:* 41–44 (1979).

32 Lee, H.W.; Lee, P.W.; Laehdvirta, J.; Brummer-Korvenkontio, M.: Aetiological relation between Korean hemorrhagic fever and nephropathia epidemica. Lancet *i:* 186–187 (1979).

33 Lee, H.W.; Lee, P.W.; Seong, I.W.; Park, D.S.: Blood analysis of normal and infected *Apodemus agrarius* with KHF virus, natural reservoir of Korean hemorrhagic fever. Korean J. Virol. *8:* 21–27 (1979).

34 Lee, H.W.; Paik, K.Z.; Seong, I.W.; Baek, L.J.: Epidemiologic study of localized outbreak of Korean hemorrhagic fever in Kyungido Tongduchun, Songnaeri. Korean J. Virol. *9:* 1–6 (1979).

35 Lee, H.W.; Lee, M.C.; Cho, K.S.: Management of Korean hemorrhagic fever. Pract. Therap. *2:* 15–21 (1980).

36 Lee, H.W.; Lee, P.W.; Baek, L.J.; Won, D.S.; Kim, W.D.; Cho, B.Y.; Lee, M.C.: Korean hemorrhagic fever. IV. Serological diagnosis. Korean J. Virol. *10:* 7–13 (1980).

37 Lee, H.W.; Seong, I.W.; Won, D.S.; Kim, W.D.; Cho, B.Y.; Lee, M.C.: Korean hemorrhagic fever. V. Complement, immunoglobulins and serum proteins. Korean J. Virol. *10:* 15–26 (1980).

38 Lee, H.W.; Bark, D.H.; Baek, L.J.; Choi, K.S.; Whang, Y.N., Woo, M.S.: Korean hemorrhagic fever patients in urban areas of Seoul. Korean J. Virol. *10:* 1–6 (1980).

39 Lee, H.W.; Doo, C.D.; Baek, L.J.: The study on breeding season of *Apodemus agrarius*, the natural host of Korean hemorrhagic fever in Korea. Korean J. Virol. *11:* 1–5 (1981).

40 Lee, H.W.; French, G.R.; Lee, P.W.; Baek, L.J.; Tsuchiya, K.; Foulke, R.S.: Observations on natural and laboratory infection of rodents with the etiologic agent of Korean hemorrhagic fever. Am. J. trop. Med. Hyg. *30:* 477–482 (1981).

41 Lee, H.W.; Lee, P.W.; Baek, L.J.; Song, C.K.; Seong, I.W.: Intraspecific transmission of Hantaan virus, the etiologic agent of Korean hemorrhagic fever, in the rodent *Apodemus agrarius*. Am. J. trop. Med. Hyg. *30:* 1106–1112 (1981).

42 Lee, H.W.; Baek, L.J.; Miyamoto, H.; Kawamata, J.; Yamanouchi, T.: Serologic diagnosis of patients with Songo fever which occurred 34 years ago in China with Hantaan virus, the etiologic agent of Korean hemorrhagic fever (in preparation).

43 Lee, P.W.; Gajdusek, D.C.; Gibbs, C.J.; Xu, Z.Y.: Aetiological relation between Korean hemorrhagic fever and epidemic hemorrhagic fever with renal syndrome in People's Republic of China. Lancet *i:* 819–820 (1980).

44 Lee, P.W.; Gibbs, C.J.; Gajdusek, D.C.; Hsiang, C.M.; Hsiung, C.D.: Identification of epidemic hemorrhagic fever with renal syndrome in China with Korean hemorrhagic fever. Lancet *i:* 1025 (1980).

45 Myhrman, G.: Nephropathia epidemica, a new infectious disease in Northern Scandinavia. Acta med. scand. *140:* 52–56 (1951).

46 Sheedy, J.A.; Froeb, H.F.; Batson, H.A.; Conley, C.G.; Murphy, J.P.; Hunter, R.B.; Gugell, D.W.; Giles, R.B.; Bershadsky, S.C.; Vester, J.W.; Yoe, R.H.: The clinical course of epidemic hemorrhagic fever. Am. J. Med. *16:* 619–628 (1954).

47 Smadel, J.E.: Epidemic hemorrhagic fever. Am. J. publ. Hlth *43:* 1327–1330 (1951).

48 Smorodintsev, A.A.; Dunaevskii, M.I.; Kakhreidze, K.A.; Neustroev, V.D.; Churilov, A.U.: Etiology and clinics of hemorrhagic nephroso-nephritis. Moscow Medgiz. *1944:* 26–47.

49 Smorodintsev, A.A.; Kazbintsev, L.I.; Chumakov, V.G.: Virus hemorrhagic fevers, Gimiz Gosndrastvennse Isdatel'stro Meditsinskoi Literatury, Leningrad, 1963 (Israel program for Scientific Translation, Jerusalem 1964), pp. 19–25 (Clearing House, Springfield, Va. 1964).

50 Svedmyr, A.H.; Lee, H.W.; Berglund, A.; Hoorn, B.; Nystrom, K.; Gajdusek, D.C.: Epidemic nephropathy in Scandinavia is related to Korean hemorrhagic fever (letter). Lancet *i:* 100 (1979).
51 Svedmyr, A.; Lee, P.W.; Gajdusek, D.C.; Gibbs, C.J.; Nystrom, K.: Antigenic differentiation of the viruses causing Korean hemorrhagic fever and epidemic (endemic) nephropathy of Scandinavia (Letter). Lancet *ii:* 315–316 (1980).
52 Tamura, M.: Occurrence of epidemic hemorrhagic fever in Osaka City: first cases found in Japan with characteristic feature of marked proteinuria. Biken's J. *7:* 79–94 (1964).
53 Traub, R.; Hertig, M.; Lawrence, W.H.; Harris, T.T.: Potential vectors and reservoirs of hemorrhagic fever in Korea. Am. J. Hyg. *59:* 291–305 (1954).
54 Traub, R.; Wisseman, C.L.: Korean hemorrhagic fever. J. infect. Dis. *158:* 267–272 (1978).
55 Trencseni, T.; Keleti, B.: Clinical aspects and epidemiology of hemorrhagic fever with renal syndrome. Analysis of clinical and epidemiological experiences in Hungary, pp. 15–42 (Akademiai Kiado, Budapest 1971).
56 Umenai, T.; Lee, H.W.; Lee, P.W.; Saito, T.; Toyoda, T.; Hongo, M.; Hoshinaga, K.; Nobunaga, T.; Horiuchi, T.; Ishida, N.: Korean hemorrhagic fever in staff in an animal laboratory. Lancet *i:* 1314–1316 (1979).
57 Vasyta, Y.S.: The epidemiology of hemorrhagic fever with renal syndrome in the R.S.F.S.R. Zh. Microbiol. Epidemiol. Immunobiol. *32:* 49–56 (1961).

Dr. H.W. Lee, WHO Collaborating Centre for Research on Korean Haemorrhagic Fever, The Institute for Viral Diseases, Korea University, 4 Myungyun-Dong 2 Ga, Chongno-Ku, Seoul 110 (Korea)

Prog. med. Virol., vol. 28, pp. 114–144 (Karger, Basel 1982)

Cell Growth Transformation by Herpes Simplex Virus

David M. Knipe

Department of Microbiology and Molecular Genetics, Harvard Medical School, Boston, Mass., USA

Contents

I. Introduction

The herpesviruses are large nuclear DNA viruses that have been isolated from a variety of host organisms ranging from fish to humans. The herpes virion is defined as having a DNA core surrounded by three concentric layers, an icosahedral capsid covered by a tegument layer and a lipid envelope. A biological characteristic of many herpesviruses is that they can remain latent in the host organism for long periods of time in one form or another but can be reactivated periodically, leading to recurrent attacks. The severity of the secondary infection depends on various factors including the immunological status of the host.

The genomes of the herpesviruses are very large, and this is thought to in part provide coding capacity for the many gene functions needed to replicate the virus in resting cells or very specialized cells. One function that is of general interest with viruses that remain in the host organism for long periods of time is the potential oncogenicity of the virus. Several species of herpesviruses have been found in humans, and considerable study has been devoted to the study of these human herpesviruses and their possible oncogenicity. Transformation of cells in culture has proven to be a useful test for the potential oncogenicity of other DNA tumor viruses, the papovaviruses and adenoviruses. The purpose of this article is to review the studies of the ability of one common herpesvirus, herpes simplex virus (HSV), to alter the growth properties of cells in culture and the possible mechanisms for these conversions. I will review briefly the studies of transformation by other herpesviruses, especially the other human herpesviruses, in order to compare and contrast these viruses with HSV. This article will not attempt to exhaustively review the large amount of work with the other viruses, but more extensive reviews will be cited.

II. Herpes Simplex Virus

Herpes simplex virus (HSV) is a common human herpesvirus that causes cold sores or fever blisters (serotype 1 or HSV-1) or a genital infection (serotype 2 or HSV-2). Following the primary infection at the site of introduction, the virus can remain latent in the individual, probably in the sensory ganglion of the nerve fiber that innervates the site of introduction. Reactivation of the virus can then periodically occur, leading to recurrent infections and lesions at the site of introduction. Thus, in its host organism, HSV is maintained in a latent state.

In cell cultures, HSV is an extremely lytic virus. Inoculation of cells in culture with HSV-1 or HSV-2 almost always leads to a cytocidal, productive infection that yields few, if any, surviving cells. This extreme difference between the behavior of the virus in cell culture and in a host organism remains to be adequately explained. It appears likely that the virus replication during the primary infection resembles the lytic infection by the virus but that this is checked by the host immune response. The virus is retained in the sensory ganglia of the host organism in a latent state suppressed by cell-mediated immunity, but lapses in this immune response lead to activation of the virus. We then can envision two extreme states of the virus: (1) the latent state in which there may be limited gene expression but no replication and (2) the lytic state in the lesion. It is not known how many abortive events occur in a host organism during the immune response suppression of an HSV infection.

It seems apparent that if HSV does cause an oncogenic event, it must do so by either an aborted infectious cycle or a defective genome. It does not alter the resting state of its host neuron and thus its 'genetic strategy' does not involve changing the growth status of its host cell. This is emphasized by the fact that it encodes its own thymidine kinase, DNA polymerase and possibly ribonucleotide reductase. Thus, the virus would not seem to need to stimulate the host cell into a growing state in order to replicate.

A correlation has been observed between the occurrence of cervical carcinoma and the presence of antibodies to HSV-2 virion and viral specific antigens, suggesting a link between HSV-2 genital infection and cervical carcinoma [1–6]. *Rawls* et al. [7] have recently reviewed this area of study. Several difficulties have made it impossible to define HSV-2 as a definite causative agent of cervical carcinoma, either by itself or in conjunction with other agents: (1) the problem of proving that a rare event - i.e., induction of the tumor – is linked to or caused by the more frequent HSV-2 infection and (2) the problem of defining adequate control groups of individuals that differ only in their history of HSV infection. Therefore, there has been increasing interest in the effects of HSV on the host cell and the ability of the virus to alter the growth properties of the cells.

A. Structure of the HSV Genome

The genomes of HSV-1 and HSV-2 are linear double-stranded DNA, approximately 100×10^6 in molecular weight. The DNA consists of two covalently linked components, the L (long or 83×10^6 in molecular weight) and S (short or 17×10^6 in molecular weight) components. Each is composed of unique sequences bounded by inverted repeated sequences. Because of the inverted sequences, each component can invert relative to the other. Therefore, virion DNA contains the four isomers with the different orientations of the two components. One orientation of the L and S components has been chosen as the prototype (P) form. The other forms, I_L (inverted L), I_S (inverted S) and I_{SL} (inverted S and L) are generated by inversion of the two components from the P arrangement. By convention, 0 to 0.83 fractional map units represent the L component, and 0.83 to 1.0 map units represent the S component. The structure and replication of HSV DNA has been extensively reviewed by *Roizman* [8].

B. Replication of HSV in Cell Culture

As stated above, HSV is extremely lytic in cultured cells. It inhibits cellular DNA and protein synthesis [9, 10] and modifies cellular RNA synthesis [11]. Three kinetic groups of viral proteins are synthesized in HSV-infected cells. The alpha or immediate early class is expressed immediately after infection and utilizes almost entirely host gene products for their expression. The alpha proteins are involved in promoting later viral gene product expression. The next class of products, the beta proteins, are largely involved in replicating the viral DNA and promoting expression of the gamma or late viral proteins. The gamma proteins are virion structural proteins and are used to assemble progeny virions. The HSV replication cycle has been reviewed in detail by *Spear and Roizman* [12].

Several techniques have been used to map the locations of the viral genes on the HSV genome. These include intertypic recombinant analysis [13–19], hybrid selection and translation of mRNA [19, 20], and electrophoretic separation and hybridization of viral mRNAs [21–26]. This large amount of molecular biology can be summarized by stating that genes encoding alpha polypeptides are mapped within or near the inverted repeat regions of the L and S components while the beta and gamma coding sequences are intermixed in the unique regions of the L and S components.

C. Growth Transformation of Cells by HSV Virions

Transformation of cells involves the stable alteration of the phenotypic properties of a cell, often by the acquisition of new genetic information [27].

Two types of transformation have been described by the addition of HSV genetic information to cells in culture. The first type, termed morphologic or growth transformation, involves the conversion of a normal cell to one with altered growth patterns which parallel those of a neoplastic cell. The growth-transformed state is assayed in cell culture by altered morphology, growth in semi-solid media, growth in media containing low concentrations of serum, or escape from senescence in the case of embryonic cells. Cells isolated from these assays often are oncogenic when injected into syngeneic animals. The second type, termed biochemical transformation, involves the conversion of a cell unable to express the enzyme thymidine kinase (tk^-) to the tk^+ state by the acquisition of the HSV gene encoding the viral tk enzyme.

Because of the cytocidal nature of HSV, it was necessary to inactivate the infectivity of virions in order to demonstrate their morphological transforming ability [28]. This was done by UV irradiation [28–31] or by incorporation of neutral red dye into virions followed by exposure to visible light [32]. Alternatively, the incubation of cells infected with wild-type or temperature-sensitive mutants at a nonpermissive temperature allowed the observation and isolation of transformed cell foci [33–37]. These assays generally involved the infection of primary cells and the selection of cells that survived the normal primary cell crisis of senescence.

In general, the efficiency of transformation of cells by HSV virions has been low. *Duff and Rapp* [28, 30] infected Syrian hamster embryonic cells with UV-irradiated HSV and observed at most one transformed focus at 30 days post-infection from each 5×10^6 infected cells. *Macnab* [35] observed approximately one transformed focus per 10^6 rat embryo cells infected at the nonpermissive temperature with temperature-sensitive mutants of HSV. *Kimura* et al. [34] isolated a number of 'foci' from infected and uninfected hamster embryo cells, but only 1–2 foci from each 2×10^6 infected cells grew into continuous lines. No foci from uninfected cells grew into continuous cell lines. It is worthwhile to compare the efficiency of transformation by other DNA viruses. Rodent cells infected with polyoma virus and assayed by altered colony morphology show 1–5% of the infected cells forming transformed foci while assay by colony formation in semi-solid medium shows up to 5,000 colonies per 10^5 infected cells when counted at 7–10 days post-infection [38]. This is a nonpermissive host cell for polyoma virus and thus cannot be directly compared to infections by UV-irradiated HSV, but the difference in efficiency is apparent. Adenovirus transformation of hamster embryo cells by infectious virions results in approximately 10 foci per 10^6 infected cells [39]. This is also a nonpermissive host cell system, but is only about 10- to 20-

fold higher than HSV. However, UV treatment of Ad12 virus enhances transformation of hamster embryo cells and raises the efficiency to approximately 30 foci per 10^6 cells [40]. These latter reports from the Ad12 system are in fact the procedures on which the experiments of *Kimura* et al. [34] were based. Therefore, transformation by HSV virions is relatively inefficient, but it is not known whether this is due to an inefficient transforming function or to the cytotoxicity of the virions.

The HSV-1 and HSV-2 transformed cell lines isolated by *Duff and Rapp* [29, 30] contained viral antigens as assayed by immunofluorescence with hamster anti-HSV serum and fluorescent anti-hamster serum. The cells therefore appeared to retain the viral coding sequences for these viral proteins. *Collard* et al. [41] found that the 333-8-9 HSV-2 transformed cell line of *Duff and Rapp* expressed RNA complementary to about 11% of the viral genome. *Frenkel* et al. [42] found that 8–32% of the HSV-2 genome was present in various transformed cell lines and tumor cell lines with lower values for higher passage cell lines. *Minson* et al. [43] detected approximately one copy of 40% of the HSV-2 genome in the 333-8-9 cells at early passages after receipt of the cells. However, later passages failed to show any detectable HSV-2 DNA sequences by liquid hybridization [43] although some subclones were considered to be positive for viral DNA and RNA sequences by in situ hybridization [44]. It is therefore unclear whether the inability to detect HSV nucleic acids in later passages is due to a complete loss of the viral sequences due to instability or an inability to detect a small remaining fragment of viral DNA. It is certain that viral DNA sequences are lost during passage of these cells, but the early passage cells had not been cloned and thus could be heterogeneous in their viral DNA content. Some investigators have been unable to detect any viral DNA in these cell lines [45]. It is not known whether these cells had lost all viral DNA or whether it could not be detected because of the design of the hybridization experiments [46].

It is clear that the use of full-sized viral DNA as a hybridization probe does not constitute a very sensitive hybridization probe. *Galloway* et al. [47] have performed hybridization experiments using the 333-8-9 HSV-2 transformed cells and their clonal derivatives with individual HSV-2 DNA fragments as probes. They found that the viral sequences from positions 0.21 to 0.33 and 0.60 to 0.65 of the viral genome were maintained in nearly all of the subclones. Not quite all of the viral genome was represented in the hybridization probes; thus they cannot exclude the possibility that other portions of the genome are also maintained. They estimate that the retained viral sequences are present in at least one copy per cell. The ideal experiment at this time would be hybridizations with small viral DNA fragments cloned in a bacterial vector.

A recent study compared the properties of cells that were spontaneously transformed, transformed with UV-irradiated HSV-2, or transformed with SV40 [48]. This study used normal mouse embryonic cells from which spontaneously transformed cells could be isolated at passage 40. Infection of these cells at passage 17 with UV-irradiated HSV led to transformed foci within 1–4 passages. Infection of these cells with SV40 led to a high frequency (1–4%) of the cells forming transformed foci. However, the HSV-induced foci yielded cell lines that were very similar in their phenotype to the spontaneously transformed cells that arose at later passages. They were very different in their growth properties from the SV40-transformed foci in that they were not nearly as efficient at forming colonies in semi-solid media, were more contact-inhibited, and did not grow as well with low serum concentrations as the SV40-transformed cells. Not all of the SV40-transformed cells were tumorigenic, but the frequency of tumorigenic foci was still much higher for SV40. These investigators concluded that transformation by HSV may be analogous to spontaneous transformation, only occurring earlier in HSV-infected cells. Thus, HSV may either act as a mitogen and stimulate cell division until the normal spontaneous transformation occurs or hasten the spontaneous transformation event by some cytopathic effect of its replication. These investigators could not detect any HSV RNA in the cells by in situ hybridization; however, the sensitivity of the assay used was not determined. They suggested that their results were more consistent with a hit-and-run mechanism in which the viral replication or a viral gene product induces an altered cellular phenotype. However, once the change in growth properties has occurred, the viral function is no longer needed. This has been described as a hit-and-run event [48]. The loss of viral DNA from the HSV-transformed cells described above and in the following section could support this hypothesis if they truly contain no viral DNA.

D. Growth Transformation by Transfection of HSV DNA Fragments

Recently, DNA fragments of the HSV genome have been tested for their ability to transform cells in culture, in order to map the location of the viral transforming sequences. Some of the reported results are conflicting; however, very different assays have been used to monitor cell transformation. Therefore, in this section I will review the current knowledge about the transforming sequences in light of the methods used to assay them.

Wilkie et al. [49] first showed that transfection of embryonic rat cells with sheared HSV-1 and HSV-2 DNA molecules would result in transformed foci in treated cell cultures. Cells from the transformed foci were grown into cell lines. Cells from continuous lines continued to express viral antigens.

Camacho and Spear [50] demonstrated that the *Xba*I-F fragment from map positions 0.30 to 0.45 from HSV-1 strain F DNA could transform cells. They used embryonic hamster cells and selected colonies that could grow in 1% serum. Thus, the selection was both for ability to grow in low serum and for immortalization of embryonic cells. They found colonies almost exclusively in the cultures transfected with the *Xba*I-F DNA fragment. The resulting clonal lines of cells could form colonies in semi-solid media, could grow in low levels of serum, and grew to higher saturation densities than control cells. The early-passage cells expressed an antigen that reacted with anti-glycoprotein A/B serum. These glycoproteins are encoded within the *Xba*I-F DNA fragment [51], and this observation was consistent with the introduction of this viral DNA fragment into the cells. However, later-passage cells contain no detectable viral DNA [52], although the assay used might not have detected a small insertion of viral DNA.

Reyes et al. [53] also mapped the transforming sequences of HSV-1 within the *Bgl*II-I DNA fragment of HSV-1 strains MP and STH2 (HFEM) or map positions 0.31 to 0.42 in the middle of the L component. They have termed this the *mtr*I region for morphological transformation of HSV-1 to distinguish it from the transforming DNA sequences of HSV-2 or *mtr*II. This was necessary because, interestingly, the transforming DNA fragment of HSV-2 DNA was not colinear with that of HSV-1 DNA. They found that the *Bgl*II-N fragment of HSV-2 strain 333 from map positions 0.58 to 0.62 would transform cells. These results were obtained in experiments that used focus formation by NIH mouse 3T3 cells as the assay. Thus, they selected for a change in growth properties, not immortalization as with many of the other HSV transformation assays. These cells do, however, undergo spontaneous transformation. With the HSV-1 DNA fragment these investigators observed 5–10 foci per plate which were visible at 3 or 10 weeks post-transfection in separate experiments. With HSV-2 DNA fragments, they observed at 3–10 weeks approximately 3 foci per plate or an efficiency of 6 foci/μg of *Bgl*II-N DNA fragment.

Transformed cell lines derived from transfection of the HSV-1 fragment contained no detectable sequences (less than 0.1 copy per cell) from the *mtr*I fragment. Viral DNA sequences were observed in the cells transformed with the HSV-2 DNA fragment only when the cellular DNA was separated by reversed-phase chromatography prior to gel electrophoresis and blot hybridization of cellular DNA. Thus, some viral DNA sequences may be retained although the exact number of copies is difficult to estimate. It seems likely that the viral DNA sequences were present in less than one copy per cell.

Thus, the transforming regions of HSV-1 and HSV-2 DNAs do not appear to be colinear. *Reyes* et al. [53] could not detect any cross-homology between *mtr*I and *mtr*II DNA fragments that would suggest rearranged sequences between the genomes of the two serotypes. Furthermore, they suggested that the *mtr*II sequences may be absent from the HSV-1 genome because the HSV-1 genome appears to lack sequences in the 0.58 to 0.62 map unit region of the HSV-2 genome.

Jariwalla et al. [54] found that low levels of unsheared viral DNA can cause transformation when transfected into Syrian hamster embryo cells. Presumably, infectivity is inactivated by shearing or nuclease activity prior to or after entry into the cells in this situation. Following transfection, the cultures were serially passaged and at 5–6 passages all cultures entered a crisis period. Cells in the uninfected cultures entered senescence 2–3 weeks later. 1–3 weeks later transformed cells overgrew the culture in infected cultures. These cells showed altered growth properties and were tumorigenic in newborn hamsters. The cells also contained antigens that reacted with antisera to viral proteins. Thus, this assay appears to select for escape from senescence or immortalization of the embryonic cells with oncogenic transformation also resulting. The assay does not quantitatively measure transformation because the cells are passed several times before transformed cells are observed.

This assay was used to test DNA fragments from throughout the HSV-2 genome [55]. These investigators found that the *Bgl*II-C DNA fragment of HSV-2 strain S-1 (map positions 0.41 to 0.58) contained the transforming activity. This maps to the left of the *mtr*II region of *Reyes* et al. [53]; however, the assays are so different that it is difficult to resolve the difference in mapping. The S-1 strain DNA also contains a *Bgl*II profile very much altered from that of most HSV-2 strains. While HSV-2 strains vary somewhat in their DNA cleavage profile, S-1 contains at least 2–3 *Bgl*II DNA fragments altered from the standard pattern. This has rarely been observed for other natural isolates of HSV-2. Thus, S-1 may differ radically from most other HSV-2 strains in terms of its DNA structure or organization.

Recent studies have used DNA fragments of HSV-2 DNA which have been cloned and propagated in plasmid vectors. These studies have found that the cloned *Bgl*II-N fragment will cause embryonic rat cells to form transformed foci [*Cameron, Wilkie* and *Macnab,* personal commun.] or will cause primary rat or mouse 3T3 cells to form transformed foci [56]. Thus, there is substantial evidence that this region of the HSV-2 genome contains transforming sequences.

Galloway and McDougall [56] have used the cloned DNA fragment as a hybridization probe to detect the viral sequences maintained in the transformed cells. They have estimated that each cell contains approximately 0.1 copy of the DNA fragment per cell. Thus, it appears to be a general observation that cells transformed by HSV DNA do not stably maintain one copy per cell, of any detectable sequences from the putative transforming fragment.

E. Biochemical Transformation

Biochemical transformation of cells with HSV virions also requires inactivation of infectivity. UV-irradiated virions have commonly been used to transform tk^- cells to the tk^+ phenotype [57–59]. Sheared DNA has also been used to transfect tk^- cells to select for tk^+ colonies [60]. Fragments of HSV DNA have also been used for mapping of the biochemical transforming function [53, 61–64]. The tk gene has been mapped by other criteria [13, 65, 66], and the gene has been entirely sequenced [Wagner, Sharp and Summers, personal commun.; ref. 67]. It now appears that only the sequences from the presumed start-site for transcription to the 3′ end of the tk gene coding sequences are needed for biochemical transformation by transfection of DNA fragments (comparison of sequence data with data of *Colbere-Garapin* et al. [63]). That is, only the tk gene itself is needed. In these transformed cells the viral DNA fragment is retained within the cells at a level of at least one copy per cell [68–70].

Biochemical transformation with virions may require more viral gene products than the tk enzyme itself. For example, binding, penetration and uncoating of the viral genome presumably must occur. It is not known what gene products must be expressed or whether viral DNA replication is needed for stable biochemical transformation to occur. Studies of biochemical transformation have been performed with temperature-sensitive mutants of HSV-1; however, no mutants have been isolated that are temperature-sensitive for biochemical transformation except those with thermolabile tk enzymes [71, 72a]. Interestingly, *Rapp* et al. [72a] identified some temperature-sensitive mutants that are totally defective for biochemical transformation even at the permissive temperature. However, these viruses express the viral tk enzyme at both temperatures during lytic infections. These viruses may have nonlethal secondary mutations in functions that are nonessential for lytic replication but are needed for biochemical transformation with virions. These may be gene functions needed for stable association of viral DNA with cells, during either latent infection or transformation.

The efficiency of biochemical transformation is significantly higher than that for morphological transformation. The efficiency of biochemical transformation by virions is one transformed cell per 10^4 to 10^6 infected cells [58, 59] and the efficiency of biochemical transformation by transfection is one transformed cell per 10^4 or higher [61; *S. Silverstein,* personal commun.]. Several explanations could account for this difference in efficiency. First, the two different regions of the genome may have different abilities to be integrated or maintained stably in the cell. Second, the selection placed on the cells in the biochemical transformation assay may require that part of the viral genome be maintained rather than quickly lost. Third, the efficiency of phenotypic conversion may be significantly greater for biochemical transformation than for morphological transformation. For example, morphological transformation may require multiple events and thus not every event introducing the morphological transforming sequences may lead to phenotypic transformation.

F. Relation between Biochemical and Morphological Transformation

Both biochemical and morphological transformation may involve the association of HSV genetic material with the host cell. Because biochemical transformation occurs at a much higher frequency, it has often been used as an assay of the ability of an HSV genome to transform a cell [58, 59]. In some cases it was found that morphological transformation accompanied biochemical transformation [59, 72b]. This raised the question of whether there was some association between the two types of transformation. However, the recent studies mapping biochemical transformation as separate from the morphological transforming region argue that these are distinct events [53]. On the other hand, the selection of cells that have incorporated the tk gene (biochemically transformed) may enhance the identification of cells that have incorporated the morphological transforming sequences. This is clearly the case in DNA-mediated transformation [73] and may also be operative in virion-mediated transformation [59].

G. Viral Antigens in Transformed and Tumor Cells

An alternative approach to determining the viral gene functions needed for morphological transformation is the identification of viral gene products that are expressed consistently and reproducibly in transformed cells. In addition to the glycoprotein A/B antigens that have been identified in transformed cells [50; *R. Courtney*, personal commun.], several other antigens have been identified including the viral thymidine kinase [74]. *Flannery* et al. [75] have found that antiserum to a large infected-cell polypeptide (VP143) in HSV-2 strain 186-infected cells reacts with an antigen in a variety of HSV-2-transformed cell lines. Interestingly, the antigen is found in the nucleus of lytically infected cells but is spread throughout the HSV-2-transformed cells.

The amount of antigen detectable in the transformed cells was dependent on the stage of the cell cycle: transformed cells showed maximal amounts of the antigen during interphase. The VP143 protein is synthesized at maximal amounts at 2–6 h post-infection in lytic infections and thus is probably a beta protein.

Not all transformed cell lines examined showed detectable amounts of VP143, and thus the antigen may not be required for transformation. However, preliminary data show a correlation between the amount of the VP143 antigen in transformed cells and the oncogenicity of the cells when injected into newborn hamsters. It has been observed that the more antigen-positive cells in a cell line, the more oncogenic the cell lines have proven to be [unpublished data cited in ref. 75]. Thus, there may be a link between the level of expression of this antigen and oncogenicity.

Dreesman et al. [76] have prepared antisera to two HSV-2 DNA-binding proteins, named ICSP 11/12 and ICSP 34/35. ICSP 11/12 is 132,000 in molecular weight and probably corresponds to ICP8 in the *Honess and Roizman* nomenclature and to VP132 in the *Marsden and Subak-Sharpe* nomenclature [77, 78]. The analogs to ICSP 34/35 in the other nomenclatures are difficult to identify. These antibodies react with transformed cells and cervical carcinoma tissue sections [76]. Like the VP143 antigen described above, the ICSP 11/12 antigen is present in the nucleus of productively infected cells but distributed throughout tumor and transformed cells. It is interesting to note that *Dreesman* et al. [76] report that ICSP 11/12 comigrates with VP143 in their gel system. Therefore, the two proteins may be similar in size in spite of the difference in their reported molecular weight. The DNA-binding protein, ICP8, or VP132, has been mapped by intertypic recombinants [13, 14, 19] and hybrid selection of mRNA [19] to map positions 0.38 to 0.41 in the L components. This is contained within the *mtr*I region but not within the *mtr*II region.

An antigen named AG-4 has been reported from *Aurelian*'s laboratory [79, 80] to correlate with onset of cervical tumors. This antigen is thought to correlate with a 161,000-molecular-weight virion envelope protein (ICP10; [ref. 81]) although antisera made to ICP10 may have been made to a mixture of proteins. The analogous polypeptide from other studies of virion proteins is not readily apparent. An antigen that reacts with AG-4 antiserum is expressed in cells transformed by fragments of HSV-2 DNA [55].

Several other studies have identified HSV antigens expressed in HSV-transformed cells. *Gupta and Rapp* [82] immunoprecipitated surface-labeled proteins from HSV-transformed cells with tumor antisera prepared in rabbits. They observed three viral-specific bands on a polyacrylamide gel. These polypeptides do not correspond to the major infected cell surface glycoproteins, and thus their identity is unknown. Two reports have demonstrated HSV type-common antigens in transformed cells [83, 84], one of which is probably related to glycoprotein D [83]. Several reports have shown that sera from animals bearing tumors induced by HSV-transformed cells contain antibodies that neutralize viral infectivity [29, 30, 34, 85, 86]. This last result may only mean that the transformed cells expressed these antigens at the time of introduction into the animal. The tumor cells may not continue to express these antigens.

In spite of these extensive studies, no viral antigen appears to be common to all HSV-transformed cells. Furthermore, no nuclear T-antigen (tumor antigen) analogous to that of the papovaviruses and adenoviruses has been identified. One major deficit in many of these antigen studies is that the transformed cell antigen has not been shown to be biochemically similar in any way to a viral-specific protein expressed in lytic infections.

It should be noted that there is no definitive evidence that HSV transformation requires viral proteins or gene products for the initiation or maintenance of growth transformation. It could occur by the insertion of viral DNA sequences such as promoter sequences into the cellular genome [87, 88]. Although viral isolates differ in their transforming ability, no viral mutants have been identified that are altered in their ability to induce growth transformation. In fact, there has been no systematic study reported that examines the ability of HSV temperature-sensitive mutants to growth-transform cells at the permissive and nonpermissive temperatures. The only reported data concerning the ability of ts mutants to transform cells examine only the events at the nonpermissive temperature [33–35]. It will be necessary to identify ts mutants that are temperature-sensitive for their ability to transform cells in order to prove that viral proteins are needed for transformation. This type of approach may be hindered by the low frequency of transformation by the wild-type virus. That is, due to the low frequency of transformation by HSV, it may be difficult to definitively identify mutants that are deficient in transformation.

H. Viral Nucleic Acids in Transformed and Tumor Cells

In addition to those reports described above concerning the presence of viral DNA in transformed cells, other studies have demonstrated the presence of viral nucleic acids in tumor cells and transformed cells. *Frenkel* et al. [89] demonstrated viral RNA sequences and a portion of the viral DNA sequences in human cervical carcinoma tissue. More recently, *McDougall and Galloway* [90] have reported data from in situ hybridization experiments demonstrating that viral-specific RNA is expressed in metastatic tumor tissue.

Two genetic approaches have been used to detect the presence of viral genetic information that has been maintained in transformed cells. These involved (1) the complementation of replication of super-infecting temperature-sensitive mutants at the nonpermissive temperature, and (2) the rescue of genetic information by recombination between the resident viral DNA sequences and the superinfecting ts mutant genome leading to a wild-type virus.

Kimura et al. [91] showed that the replication of two HSV-2 ts mutants, tsG4 and tsD6, could be complemented for replication at the nonpermissive temperature when grown in HSV-2-transformed cells. Thus, these two viral genes may be expressed or retained in these transformed cells. The defective gene functions are not known; however, the tsG4 lesion has been mapped within the HSV-2 *Hin*dIII-E DNA fragment or map positions 0.42 to 0.50 map units [*D. Knipe*, unpublished observations].

Macnab and Timbury [92] demonstrated that three mutants of nine tested, ts10, ts3, and ts4, were complemented for replication when grown at the nonpermissive temperature in HSV-2-transformed cells. Thus, several viral gene functions may be retained or expressed in these cells. No wild-type recombinants were detected in the progeny from these infections. A later study of the progeny from superinfection of these same transformed cells has shown that some of the progeny are wild-type or ts^+ [93]. In these experiments the superinfecting virus was of a different serotype from the original transforming virus. Therefore, the DNA genome of the recombinants could be analyzed to map the recombination events. Most of the resulting recombinants showed DNA structures consistent with intertypic recombinational events resulting from recombination between the resident transforming genome and the superinfecting genome. However, a few progeny viruses showed DNA profiles that were identical to the original transforming genome. These presumably contained some genetic sequences altered from the original ts transforming virus because the phenotype of the progeny virus was ts^+. Thus, the progeny virus contained either an undetectable cross-over or a reverting mutation, a situation not uncommon when intertypic recombinants generated by marker rescue are analyzed. At any rate, it appears that most or all of the original transforming genome has been maintained for many generations in the transformed cells.

I. Effects of HSV on the Host Cell

The above discussion presents the current knowledge concerning the identification of the transforming sequences of HSV DNA. With the large number of gene products encoded by the HSV genome many possible cell alterations could be envisioned that would lead to altered growth properties. Some of the effects of HSV on the host cell that could lead to altered growth are as follows.

1. Chromosome Damage.

Numerous early reports demonstrated that HSV could induce chromosomal damage to the host cell during an abortive or productive infection [94–99]. Some chromosomal changes are also apparent in HSV-transformed cells [100]. Some of the changes that have been observed are secondary constrictions, single chromatid breaks, multiple chromatid breaks, complete fragmentation, chromatid exchange and centromeric fusions. In at least some cell lines, viral replication does not need to occur in order for chromosomal damage to occur. Ultraviolet irradiation of virions inactivates their infectivity 4 times faster than their ability to cause chromosomal damage [98]. Thus, it is thought that an early gene product is responsible. In light of the genome inversions and site-specific recombination shown to occur during HSV DNA replication [8, 17, 101], it would not be surprising if the cellular DNA were affected also. This needs to be considered as a mechanism of transformation in light of *Cairns'* [102] recent hypothesis that most human cancers are the result of genetic transposition.

2. Stimulation of Cellular DNA Replication

The papovaviruses stimulate host cell DNA replication as an early step in their lytic cycle and transformation events. In contrast, HSV usually inhibits host cell DNA replication early in its lytic cycle as stated above. However, *Yamanishi* et al. [103] found that cellular DNA replication in serum-starved cells was stimulated by infection at 38.5 °C with an HSV-2 ts mutant, ts4. This mutant can also transform cells at the nonpermissive temperature. *Melvin and Kucera* [104] found that viral replication is inhibited at 42 °C. Furthermore, HSV infection of serum-starved cells at 42 °C stimulates cellular DNA replication. It appears, therefore, that if viral replication is blocked in a resting cell, HSV-2 can under some unusual conditions stimulate cellular DNA replication rather than inhibit it as during the usual productive infection [9].

3. Alterations of the Host Cell Cytoskeleton

One of the most dramatic effects of HSV infection is the change in the cell shape. Most natural isolates of HSV cause cells to round up and form refractile spheres following infection [105]. The cells detach from the substratum much later in the infectious cycle. It is not known why the infected cell changes shape so drastically, but it seems likely that infection affects the cytoskeleton. This could result in altered cell growth properties if the cell is not killed by the infection.

4. Activation of RNA Tumor Virus Genomes

It has been well documented that infection with inactivated HSV can activate endogenous RNA tumor viruses from some cell lines [106–108]. It was therefore suggested that HSV may actually transform cells by activation of another virus. This interesting hypothesis has been difficult to prove or disprove. Evidence of retroviruses has been found in some HSV-transformed cell lines but not in others [29, 30, 56]. Perhaps the most informative results on this matter have come from experiments performed to map this viral function [109]. It was found that several DNA fragments of the HSV-1 genome could activate RNA tumor virus production but these did not correlate with the *mtr*I sequences. Thus it appears that activation of RNA tumor viruses does not equate in a simple manner with morphological transformation by HSV-1. Instead, activation of retroviruses may occur by some other physiological perturbation of the cells.

J. Summary on HSV

In summary, infection by HSV can induce the formation of cell foci that have altered growth properties and altered morphology. HSV is relatively inefficient at morphological transformation compared to the other DNA tumor viruses. Many of the transformation assays with HSV involve the use of embryonic cells and select for transformed, immortalized cells. All of the assays employ normal cells that are capable of significant levels of spontaneous transformation. Our own attempts to transform cell lines, such as the rat F111 cell line that does not spontaneously transform, have been negative [*Knipe,* unpublished data]. It has been hypothesized by some that transformation by HSV is a 'hit-and-run' event that merely triggers or speeds up the spontaneous transformation event. In support of this idea, cells transformed by HSV often do not retain a single copy of HSV sequences in each cell, especially cells transformed by HSV DNA. It is not yet clear whether or not all HSV DNA sequences have been lost by these cells.

In light of the biology of HSV infections, it might not be surprising that cell transformation by HSV is not equivalent to that by papovaviruses, for example. Papovaviruses need to stimulate host cell replication and DNA synthesis to replicate their own DNA because they do not encode a DNA polymerase or the nucleotide metabolism enzymes needed for their DNA replication. On the other hand, HSV encodes its own DNA polymerase, pyrimidine kinase and possibly ribonucleotide reductase. It thus could and does remain latent and replicate in resting cells. Under some circumstances HSV may induce host cell DNA replication but this may not be universal. Transformation by HSV thus may not be due to a molecular event that is a direct part of the replication cycle, but rather, may be a side product of it.

III. Epstein-Barr Virus

Epstein-Barr virus (EBV) is a second common human herpesvirus that infects most individuals at some time during their lifetime. Infection of children results only in mild disease while infections of adults lead to infectious mononucleosis. B lymphocytes isolated from most individuals are latently infected with EBV. The virus has also been associated with nasopharyngeal carcinoma and Burkitt's lymphoma, malignancies occurring predominantly in southern China and Africa, respectively. It is not known how the disease associated with the virus can be so different in different geographical regions.

The study of the structure of virion DNA and DNA fragments cloned in prokaryotic vectors has allowed a detailed analysis of the genome structure [110–117]. These studies have shown that the DNA is linear, double-stranded, and approximately 115×10^6 in molecular weight. A tandem direct repeat of 0.5 kilobase pairs (kbp) is present in several copies at both ends of the molecule. The internal unique sequences are divided into two regions (L and S) by internal tandem direct repeats of a 3.3 kbp sequence. No inversions of these regions occur.

HSV and EBV have been called a 'study in contrasts' because the two classes of viruses are so different in their biological properties [46]. The biological properties of EBV have been recently reviewed by *Miller* [118]. Unlike HSV, EBV does not produce a lytic infection in vitro. B lymphocytes are the only cells that can be infected by EBV, but little infectious virus is produced. Many of the infected lymphocytes acquire the ability to replicate indefinitely in culture. In fact, all continuous B-lymphocyte cell lines have been found to be infected with EBV. This conversion is called growth transformation, or immortalization. Most cells in the resulting culture do not produce progeny virus; however, a small percentage of the cells do produce virus. Almost all cells contain the EBV-specified nuclear antigen, called EBNA. This antigen appears before the cellular DNA replication is stimulated, and expression of the antigen is not affected by inhibitors of DNA synthesis [119]. Virus-producing cells contain other viral antigens, the early antigen (EA) and the viral capsid antigen (VCA). These presumably represent viral proteins made later in the productive replication cycle. Some cell lines, often marmoset lymphocyte lines, produce higher quantities of virus and these have been used as a source of virus for biochemical studies. Because there is no lytic system, there is no plaque assay for EBV and knowledge of its genetics is very limited.

Quantitative assays for transformation by EBV have been established [120, 121]. In these assays, serial dilutions of virus are added to microwells containing human umbilical cord lymphocytes. Transformed cell growth is apparent in wells containing transformed cells within a few days, and cell growth is scored at 8 weeks. It was observed that approximately 1 of 50 EBV particles was capable of transformation. Serial dilution of EBV-infected lymphocytes gave a transformed center assay, and by this assay it was calculated that 0.5% of the infected cells yielded transformed colonies. However, established cell lines generally have plating efficiencies of approximately 10%. Thus, as many as 5% of the cells may be transformed. Thymidine labelling followed by autoradiography of infected cells shows that DNA synthesis is stimulated in about 3% of the cells [118]. Thus, most cells stimulated into DNA synthesis could go on to be immortalized.

The rate of inactivation of the transforming activity of EBV by irradiation of virions is approximately equal to the rate of inactivation of HSV infectivity [122]. Because the genomes of the two viruses are approximately the same size, it was concluded that most or all of the EBV genome is needed for immortalization of lymphocytes.

Most of the DNA molecules present in immortalized lymphocytes are covalently closed, circular, super-coiled molecules [112, 123, 124]. Apparently, the viral DNA circularizes using the terminal direct repeated sequences once it is within the cell. It is therefore maintained as an episomal molecule. There have also been reports that some of the molecules are integrated into the cellular DNA [125]. Due to the large number of copies of the extrachromosomal viral DNA, it is difficult to determine whether there are indeed integrated copies of EBV DNA or whether these are free DNA molecules that are trapped within the chromosomal DNA. Molecular cloning of the putative integrated EBV sequences would seem to be the best way to resolve this difficult problem.

RNA expressed in the latently infected, immortalized lymphocytes could be encoding gene functions needed for the maintenance of growth stimulation. Thus, it was important to map these RNAs on the genome. *King* et al. [126] have shown that an infected human neonatal lymphocyte cell line expresses RNA from only a region just to the right of the internal repeat region and the internal repetition itself and neighboring fragments. These results were very similar to RNA expressed by Namalwa and Raji cells from Burkitt tumor biopsies [127]. Thus, these regions may be encoding transformation functions.

Only one virus isolate is known that is defective for lymphocyte immortalization, the P3-HR1 strain. This isolate was obtained during attempts to select more permissive clones from a Burkitt tumor cell line. Analysis of the structure of the P3-HR1 DNA has shown that it has a deletion in a region near the right-hand boundary of the internal repetitive sequences [111]. This maps within the region encoding RNA in latently infected lymphocytes, and thus the deletion may explain the loss of transforming capacity. However, there is other evidence that P3-HR1 DNA molecules are heterogeneous and some may comprise defective DNA molecules [128]. Thus, further studies are needed to confirm that the deleted region from P3-HR1 is needed for lymphocyte transformation. The location of the coding sequences for the EBNA antigen has not been mapped on the EBV genome as yet.

It has been pointed out by *Roizman* [129] and *Schwartz* [130] that EBV may only act as a mitogen and continually stimulate lymphocytes but not have true oncogenic potential itself. Rather, a malignant lymphoma may develop only as a result of immortalization or mitogenesis of lymphocytes transformed by other agents. Further work is needed to resolve this issue.

In summary, EBV can very efficiently immortalize lymphocytes or allow them to replicate indefinitely in culture. In the latently infected cells only a limited portion of the genome is expressed, even though the entire genome is retained in the cells. No good lytic infection system has been developed. Thus, in many ways EBV shows properties that are just the opposite of those displayed by herpes simplex virus.

IV. Human Cytomegalovirus

Cytomegalovirus (CMV) is a human herpesvirus that was isolated originally from cytomegalic inclusion disease tissue. The virus replicates in cell cultures although it is fairly species-specific in terms of its host cell requirements. It is also slow-growing and very cell-associated.

UV-irradiated human CMV has been shown to be capable of transformation of embryonic hamster cells [131]. One cell line was derived from the transformed clones and it contained CMV antigens in the cytoplasm and on the cell surface. The cells were tumorigenic in weanling hamsters, and serum from tumor-bearing animals contained antibody to CMV antigens. A second system for demonstrating transformation involved the establishment of a persistent infection by a human CMV isolate in human embryonic lung cells [132]. From these infected cultures, cells arose that had altered growth properties. These cells also contain a nuclear antigen that reacts with CMV antisera [133]. This antigen was weakly apparent by immunofluorescence but brightly stained by anti-complement immunofluorescence. The latter approach was also used to detect the EBNA antigen. Thus, CMV-transformed cells may contain a nuclear antigen analogous to the T-antigen of SV-40- and adenovirus-transformed cells.

V. Equine Herpesviruses

Three subgroups of herpesviruses have been isolated from horses, equine herpesvirus (EHV) types 1, 2, and 3. EHV-1, also called equine abortion virus, causes a variety of diseases including encephalitis. EHV-2 is also named equine cytomegalovirus because of its tendency to remain cell-associated and the presence of large intranuclear inclusions and cytomegaly in cells infected with this virus. EHV-3 causes a venereal disease in horses. Thus, these three subgroups bear some similarities to HSV-1, human cytomegalovirus and HSV-2, respectively.

The genome structure of EHV-1 has been most extensively studied. It contains a long segment (L) and a short segment (S). Unlike the genome of HSV, only the short segment is bracketed by inverted repeated regions that allow the inversion of the S component [134]. Two types of approaches have been used to demonstrate the transforming ability of EHV-1. In the first, UV-irradiated EHV virions were used to infect hamster embryo cells. From these cells foci with altered morphology and lack of contact inhibition grew out and were isolated. Uninfected cultures showed no foci and did not survive more than a few passages. Cell lines were grown from the foci, and the resulting cells grew with low serum and formed colonies in soft agar [135]. These cells are tumorigenic in newborn or adult syngeneic hamsters.

The second approach used virus stocks enriched for defective interfering (DI) particles to infect primary hamster embryo cells. These infected cells also gave rise to transformed foci of cells. Cell lines derived from these foci show the same properties as the cell lines described above [136]. Interestingly, in contrast to the results described above for HSV, all of these transformed cell lines retain approximately one or more copies of viral DNA sequences from 0.32 to 0.42 map units [134]. These sequences are maintained in tumor cell DNA. The viral DNA sequences are integrated into cellular DNA, and this structure has been verified by cloning of the cell-viral DNA joint sequences [*O'Callaghan,* personal commun.]. Thus, in these cells a fragment of viral DNA may be stably integrated and maintained in the transformed cells.

VI. Simian Herpesviruses

The New World monkey herpesviruses, herpesvirus saimiri (HVS) and herpesvirus ateles (HVA), were isolated from squirrel monkeys and spider monkeys, respectively. They appear to cause no disease in their natural host and remain latent in circulating lymphocytes in the host organism [137]. These two viruses cause a T-cell lymphoma when introduced into other New World monkeys. Transformed lymphoblastoid cell lines can be established with cells from animals infected with either virus. However, in vitro transformation of lymphocytes has only been achieved with HVA [138]. It cannot be performed with HVS.

Two types of viral DNA molecules are found in HVS and HVA virions. M-DNA molecules are infectious, approximately 100×10^6 in molecular weight, and have a G+C content of approximately 70%. The M-genomes consist of light-density unique-sequence DNA bounded by tandem-reiterated units of H-DNA. The H-DNA molecules are defective molecules with long tandem arrays of a repeating subunit of approximately 1 million molecular weight. Transformed lymphoblastoid cell lines have covalently closed, circular viral DNA molecules within them that presumably replicate as episomes [139].

Attenuated virus variants of HVS have been isolated by serial passage of the virus in cell cultures. These variants have lost their oncogenic potential for various New World monkeys, but appear to replicate and persist in the host animal. The DNA genomes of two independent isolates of this type have deletions in the extreme left-hand portion of the unique-sequence DNA [139]. These variants appear to contain deletions of transforming functions not needed for replication. It is not yet known what gene products are encoded in this region of the genome.

VII. General Summary

Herpes simplex virus (HSV) enters a latent state in its host organism but is extremely lytic in cultured cells. Thus HSV could cause an oncogenic transformation event only as the result of an abortive infection. If viral replication is blocked, HSV virions or DNA can transform cultures of cells although growth transformation is an infrequent event. HSV DNA sequences are often difficult to detect in the resulting cell lines. It is not known whether these cell lines continue to maintain a small undetectable piece of HSV DNA or not. HSV-transformed cells in some cases resemble spontaneously 'transformed' cells, and thus it is tempting at this point to consider whether in some growth transformation events HSV may initiate or act as a 'promoter' of spontaneous transformation. Alternatively, HSV infection could be mutagenic to the host cell. In both of these situations, the continued presence of HSV DNA sequences would not be required; the HSV transformation would be a 'hit-and-run' event.

No viral-specific antigen has been identified which is found within all HSV-transformed cells. No mutants of HSV have been isolated that are defective for growth transformation of cells. The isolation of a temperature-sensitive mutant would be essential for the demonstration that growth transformation by HSV is mediated by a viral protein gene product and not by insertion of a DNA sequence such as a promoter sequence. For these reasons and because of the inefficiency of growth transformation, our understanding of the mechanism of growth transformation by HSV lags behind that of the other DNA tumor viruses. However, because of the possible difference in transformation mechanism from the other DNA viruses and the question of the mechanism of maintenance of the viral genome in the latent state, the interactions of HSV with its host cell remain an extremely important area of study.

Growth transformations by cytomegalovirus (CMV) and equine herpesvirus (EHV) are also low-frequency events, like that by HSV. The CMV-transformed cells are not well characterized, but the cells transformed by EHV seem to retain a common viral DNA fragment in each cell. Thus, EHV-transformed cells may be different from the HSV-transformed cells.

In contrast, Epstein-Barr virus (EBV) efficiently immortalizes B lymphocytes. The complete EBV genome is maintained in these cells, and only part of the genome is expressed as RNA in nonproducing cells. In a low percentage of the cells a productive infection results in release of infectious EBV particles. Thus, immortalization of lymphocytes is a part of the replicative cycle of EBV. Lymphocyte transformation by the simian herpesviruses resembles that by EBV in certain respects, in that it is also one step in the replicative cycle of the virus.

Addendum

Adelman, Howett and Rapp (1981)[1] have recently shown that highly tumorigenic lines of HSV-2-transformed cells have high levels of plasminogen activator. Thus, there may be a correlation between the level of protease production by transformed cells and their tumorigenicity. It remains to be demonstrated whether the increased levels of plasminogen activator are the direct result of the HSV-2 infection or a secondary effect of the transformation event.

Acknowledgments

I would like to thank the many colleagues who gave me permission to cite their unpublished work or manuscripts in press. I would like to thank Dr. *Bernard Roizman* and Dr. *Cody Meissner* for their helpful comments on the manuscript. Research cited from my own laboratory is supported by NIH grant CA 26345.

[1] Adelman, S.F.; Howett, M.K., and Rapp, F.: Tumorigenicity of herpesvirus-transformed cells correlates with production of plasminogen activator. Molecular and Cellular Biology *1:* 408–417 (1981).

References

1 Nahmias, A.J.; Naib, Z.M.; Josey, W.E.: *Herpesvirus hominis* type 2 infection: association with cervical cancer and perinatal disease. Perspect. Virol. *7:* 73–89 (1970).

2 Naib, Z.M.; Nahmias, A.J.; Josey, W.E.; Kramer, J.H.: Genital herpetic infection: association with cervical dysplasia. Cancer, N.Y. *23:* 940–945 (1969).

3 Rawls, W.E.; Tompkins, W.A.; Melnick, J.L.: The association of herpesvirus type 2 and carcinoma of the uterine cervix. Am. J. Epidem. *89:* 547–554 (1969).

4 Aurelian, L.; Royston, I.; Davis, H.J.: Antibody to genital herpes simplex virus: association with cervical atypia and carcinoma *in situ.* J. natn. Cancer Inst. *45:* 455–464 (1970).

5 Nahmias, A.J.; Josey, W.E.; Naib, Z.M.; Luce, C.F.; Guest, B.A.: Antibodies to *Herpesvirus hominis* types 1 and 2 in humans. II. Women with cervical cancer. Am. J. Epidem. *91:* 547–552 (1970).

6 Royston, I.; Aurelian, L.: Immunofluorescent detection of herpesvirus antigens in exfoliated cells from human cervical carcinoma. Proc. natn. Acad. Sci. USA *67:* 204–212 (1970).

7 Rawls, W.E.; Bacchetti, S.; Graham, F.L.: Relation of herpes simplex viruses to human malignancies. Curr. Top. Microbiol. Immunol. *77:* 72–95 (1977).

8 Roizman, B.: The structure and isomerization of herpes simplex virus genomes. Cell *16:* 481–494 (1979).

9 Roizman, B.; Roane, P.R., Jr.: The multiplication of herpes simplex virus. II. The relation between protein synthesis and the duplication of viral DNA in infected HEp-2 cells. Virology *22:* 262–269 (1964).

10 Sydiskis, R.J.; Roizman, B: Polysomes and protein synthesis in cells infected with a DNA virus. Science *153:* 76–78 (1966).

11 Roizman, B.; Bachenheimer, S.L.; Wagner, E.K.; Savage, T.: Synthesis and transport of RNA in herpesvirus-infected cells. Cold Spring Harb. Symp. quant. Biol., vol. 35, pp. 753–771 (Cold Spring Harbor Laboratory, Cold Spring Harbor 1970).

12 Spear, P.; Roizman, B.: Herpes simplex viruses; in Tooze, DNA tumor viruses, pp. 615–745 (Cold Spring Harbor Laboratory, Cold Spring Harbor 1980).

13 Morse, L.S.; Pereira, L.; Roizman, B.; Schaffer, P.A.: Anatomy of herpes simplex virus (HSV) DNA. X. Mapping of viral genes by analysis of polypeptides and functions specified by HSV-1 × HSV-2 recombinants. J. Virol. *26:* 389–410 (1978).

14 Marsden, H.S.; Stow N.D.; Preston, V.G.; Timbury, M.C.; Wilkie, N.M.: Physical mapping of herpes simplex virus-induced polypeptides. J. Virol. *28:* 624–642 (1978).

15 Preston, V.G.; Davison, A.J.; Marsden, H.S.; Timbury, M.C.; Subak-Sharpe, J.H.; Wilkie, N.M.: Recombinants between herpes simplex virus types 1 and 2: analyses of genome structures and expression of immediate early polypeptides. J. Virol. *28:* 499–517 (1978).

16 Knipe, D.M.; Ruyechan, W.T.; Roizman, B.; Halliburton, I.W.: Molecular genetics of herpes simplex virus. Demonstration of regions of obligatory and non-obligatory identity within diploid regions of the genome. Proc. natn. Acad. Sci. USA *75:* 3896–3900 (1978).

17 Roizman, B.; Jacob, R.J.; Knipe, D.M.; Morse, L.S.; Ruyechan, W.T.: On the structure, functional equivalence and replication of the four arrangements of herpes simplex virus DNA. Cold Spring Harb. Symp. quant. Biol., vol. 43, pp. 809–826 (Cold Spring Harbor Laboratory, Cold Spring Harbor 1979).

18 Knipe, D.M.; Batterson, W.; Nosal, C.; Roizman, B.; Buchan, A.: Molecular genetics of

herpes simplex virus. VI. Characterization of a temperature-sensitive mutant defective in the expression of all early viral gene products. J. Virol. *38:* 539–547 (1981).

19 Conley, A.J.; Knipe, D.M.; Jones, P.C.; Roizman, B.: Molecular genetics of herpes simplex virus. VII. Characterization of a temperature-sensitive mutant produced by in vitro mutagenesis and defective in DNA synthesis and accumulation of gamma polypeptides. J. Virol. *37:* 191–206 (1981).

20 Mackem, S.; Roizman, B.: Regulation of herpesvirus macromolecular synthesis: transcription initiation sites and domains of alpha genes. Proc. natn. Acad. Sci. USA *77:* 7122–7126 (1980).

21 Jones, P.C.; Hayward, G.S.; Roizman, B.: Anatomy of herpes simplex virus DNA. VI. *α* RNA is homologous to non-contiguous sites in both L and S components of viral DNA. J. Virol. *21:* 268–276 (1977).

22 Clements, J.B.; Watson, R.J.; Wilkie, N.M.: Temporal regulation of herpes simplex virus type 1 transcription: location of transcripts on the viral genome. Cell *12:* 275–285 (1977).

23 Watson, R.J.; Preston, C.M.; Clements, J.B.: Separation and characterization of herpes simplex virus type 1 immediate-early mRNA's. J. Virol. *31:* 42–52 (1979).

24 Anderson, K.P.; Stringer, J.R.; Holland, L.E.; Wagner, E.K.: Isolation and localization of herpes simplex virus type 1 mRNA. J. Virol. *30:* 805–820 (1979).

25 Anderson, K.P.; Holland, L.E.; Gaylord, B.H.; Wagner, E.K.: Isolation and characterization of mRNA encoded by a specific region of the herpes simplex virus type 1 genome. J. Virol. *33:* 749–759 (1980).

26 Anderson, K.P.; Costa, R.H.; Holland L.E.; Wagner, E.K.: Characterization of herpes simplex virus type 1 RNA present in the absence of de novo protein synthesis. J. Virol. *34:* 9–27 (1980).

27 Pellicer, A.; Robins, D.; Wold, B.; Sweet, R.; Jackson, J.; Lowy, I.; Roberts, J.M.; Sim, G.K.; Silverstein, S.; Axel, R.: Altering genotype and phenotype by DNA-mediated gene transfer. Science *209:* 1414–1422 (1980).

28 Duff, R.; Rapp, F.: Oncogenic transformation of hamster cells after exposure to herpes simplex virus type 2. Nature new Biol. *233:* 48–50 (1971).

29 Duff, R.; Rapp, F.: Properties of hamster embryo fibroblasts transformed *in vitro* after exposure to ultraviolet-irradiated herpes simplex virus type 2. J. Virol. *8:* 469–477 (1971).

30 Duff, R.; Rapp, F.: Oncogenic transformation of hamster embryo cells after exposure to inactivated herpes simplex virus type I. J. Virol. *12:* 209–217 (1973).

31 Duff, R.; Rapp, F.: Quantitative assay for transformation of 3T3 cells by herpes simplex virus type 2. J. Virol. *15:* 490–496 (1975).

32 Rapp, F.; Li, J.H.; Jerkofsky, M.: Transformation of mammalian cells by DNA-containing viruses following photodynamic inactivation. Virology *55:* 339–346 (1973).

33 Takahashi, M.; Yamanishi, K.: Transformation of hamster embryo and human embryo cells by ts mutants of herpes simplex virus type 2. Virology *61:* 306–311 (1974).

34 Kimura, S.; Flannery, V.L.; Levy, B.; Schaffer, P.A.: Oncogenic transformation of primary hamster cells by herpes simplex virus type 2 (HSV-2) and an HSV-2 temperature-sensitive mutant. Int. J. Cancer *15:* 786–798 (1975).

35 Macnab, J.C.M.: Transformation of rat embryo cells by temperature-sensitive mutants of HSV. J. gen. Virol. *24:* 143–153 (1974).

36 Darai, G.; Munk, K.: Human embryonic lung cells abortively infected with Herpesvirus hominis type 2 show some properties of cell transformation. Nature new Biol. *241:* 268–269 (1973).

37 Braun, R.; Flugel, R.; Munk, K.: Malignant transformation of rat embryo fibroblasts by HSV types 1 and 2 at suboptimal temperature. Nature, Lond. *265:* 744–746 (1977).

38 MacPherson, I.; Montagnier, L.: Agar suspension culture for the selective assay of cells transformed by polyoma virus. Virology *23:* 291–294 (1964).

39 Schell, K.; Schmidt, M.: Adenovirus transformation of hamster embryo cells. I. Assay conditions. Arch. ges. Virusforsch. *24:* 332–341 (1968).

40 Schell, K.; Maryak, J.; Young, J.; Schmidt, M.: Adenovirus transformation of hamster embryo cells. II. Inoculation conditions. Arch. ges. Virusforsch. *24:* 342–351 (1968).

41 Collard, W.; Thornton, H.; Green, M.: Cells transformed by HSV-2 transcribe virus specific RNA sequences shared by herpes virus types 1 and 2. Nature new Biol. *243:* 264–265 (1973).

42 Frenkel, N.; Locker, H.; Cox, B.; Roizman, B.; Rapp, F.: Herpes simplex virus DNA in transformed cells: sequence complexity in five hamster cell lines and one derived hamster tumor. J. Virol. *18:* 885–893 (1976).

43 Minson, A.C.; Thouless, M.E.; Eglin, R.P.; Darby, G.: The detection of virus DNA sequences in a herpes type 2 transformed cell line (333-8-9). Int. J. Cancer *17:* 493–500 (1976).

44 Copple, C.D.; McDougall, J.K.: Clonal derivatives of a herpes type 2 transformed hamster cell line (333-8-9): cytogenetic analysis, tumorigenicity and virus sequence detection. Int. J. Cancer *17:* 501–510 (1976).

45 Davis, D.B.; Kingsbury, D.B.: Quantitation of the viral DNA present in cells transformed by UV-irradiated herpes simplex virus. J. Virol. *17:* 788–793 (1976).

46 Roizman, B.; Frenkel, N.; Kieff, E.D.; Spear, P.G.: The structure and expression of human herpesvirus DNAs in productive infection and in transformed cells. Cold Spring Harb. Conf. Cell Prolif., vol. 4, pp. 1069–1111 (1977).

47 Galloway, D.A.; Copple, C.D.; McDougall, J.K.: Analysis of viral DNA sequences in hamster cells transformed by herpes simplex virus type 2. Proc. natn. Acad. Sci. USA *77:* 880–884 (1980).

48 Hampar, B.; Boyd, A.L.; Derge, J.G.; Zweig, M.; Eader, L.; Showalter, S.D.: Comparison of properties of mouse cells transformed spontaneously by UV light-irradiated HSV or by SV40. Cancer Res. *40:* 2213–2222 (1980).

49 Wilkie, N.M.; Clements, J.B.; Macnab, J.C.; Subak-Sharpe, J.H.: The structure and biological properties of herpes simplex DNA. Cold Spring Harb. Symp. quant. Biol., vol. 39, pp. 657–666 (Cold Spring Harbor Laboratory, Cold Spring Harbor 1974).

50 Camacho, A.; Spear, P.G.: Transformation of hamster embryo fibroblasts by a specific fragment of the herpes simplex virus genome. Cell *15:* 993–1002 (1978).

51 Ruyechan, W.T.; Morse, L.S.; Knipe, D.M.; Roizman, B.: Molecular genetics of herpes simplex virus. II. Mapping of the major viral glycoproteins and of the genetic loci specifying the social behavior of infected cells. J. Virol. *29:* 677–697 (1979).

52 Leiden, J.; Frenkel, N.: Mapping of the HSV sequences in transformed cells; in Becker, Herpes virus DNA: recent studies of the viral genome (Nijhoff, The Hague 1981).

53 Reyes, G.R.; LaFemina, R.; Hayward, S.D.; Hayward, G.S.: Morphological transformation by DNA fragments of human herpesviruses: evidence for two distinct transforming regions in HSV types 1 and 2 and lack of correlation with biochemical transfer of the thymidine kinase gene. Cold Spring Harb. Symp. quant. Biol., vol. 44, pp. 629–641 (Cold Spring Harbor Laboratory, Cold Spring Harbor 1980).

54 Jariwalla, R.J.; Aurelian, L.; Ts'O, P.O.P.: Neoplastic transformation of cultured Syrian hamster embryo cells by DNA of herpes simplex virus type 2. J. Virol. *30:* 404–409 (1979).

55 Jariwalla, R.J.; Aurelian, L.; Ts'O, P.O.P.: Tumorigenic transformation induced by a specific fragment of DNA from herpes simplex virus type 2. Proc. natn. Acad. Sci. USA *77:* 2279–2283 (1980).

56 Galloway, D.A.; McDougall, J.K.: Transformation of rodent cells by a cloned DNA fragment of herpes simplex virus type 2. J. Virol. *38:* 749–760 (1981).

57 Munyon, W.; Kraiselburd, E.; Davis, D.; Mann, J.: Transfer of thymidine kinase to thymidine kinaseless L cells by infection with ultraviolet-irradiated herpes simplex virus. J. Virol. *7:* 813–820 (1971).

58 Rapp, F.; Buss, E.R.: Comparison of herpes simplex isolates using a quantitative selection assay for transformation. Intervirology *6:* 72–82 (1975/76).

59 Rapp, F.; Turner, N.: Biochemical transformation of mouse cells by herpes simplex virus types 1 and 2: comparison of different methods of inactivation of viruses. Archs. Virol. *56:* 77–87 (1978).

60 Bacchetti, S.; Graham, F.L.: Transfer of the gene for thymidine kinase to thymidine kinase-deficient human cells by means of purified herpes simplex viral DNA. Proc. natn. Acad. Sci. USA *74:* 1590–1594 (1977).

61 Wigler, M.; Silverstein, S.; Lee, L.; Pellicer, A.; Cheng, Y.; Axel, R.: Transfer of a purified thymidine kinase gene to cultured mouse cells. Cell *11:* 223–232 (1977).

62 Maitland, N.J.; McDougall, J.K.: Biochemical transformation of mouse cells by fragments of herpes simplex virus DNA. Cell *11:* 233–241 (1977).

63 Colbere-Garapin, F.; Chousterman, S.; Horodniceanu, F.; Kourilsky, P.; Garapin, A.-C.: Cloning of the active thymidine kinase gene of herpes simplex virus type 1 in *Escherichia coli* K-12. Proc. natn. Acad. Sci. USA *76:* 3755–3759 (1979).

64 Wilkie, N.M.; Clements, J.B.; Boll, W.; Mantei, N.; Lonsdale, D.; Weissman, C.: Hybrid plasmids containing an active thymidine kinase gene of herpes simplex virus 1. Nucl. Acids Res. *7:* 859–877 (1979).

65 Stow, N.D.; Subak-Sharpe, J.H.; Wilkie, N.M.: Physical mapping of herpes simplex virus type 1 mutations by marker rescue. J. Virol. *28:* 182–192 (1978).

66 Halliburton, I.W.; Morse, L.S.; Roizman, B.; Quinn, K.E.: Mapping of the thymidine kinase genes of type 1 and type 2 herpesviruses using intertypic recombinants. J. gen. Virol. *49:* 235–254 (1980).

67 McKnight, S.L.: The nucleotide sequence and transcript map of the HSV thymidine kinase gene. Nucl. Acids Res. *24:* 5949–5964 (1980).

68 Leiden, J.M.; Frenkel, N.; Rapp, F.: Identification of the herpes simplex virus DNA sequences present in six herpes simplex virus thymidine kinase-transformed mouse cell lines. J. Virol. *33:* 272–285 (1980).

69 Kraiselburd, E.; Gage, L.P.; Weissbach, A.: Presence of a herpes simplex virus DNA fragment in an L-cell clone obtained after infection with irradiated herpes simplex virus. J. molec. Biol. *97:* 533–542 (1975).

70 Sugino, W.M.; Chadha, K.C.; Kingsbury, D.T.: Quantification of the herpes simplex virus DNA present in biochemically transformed mouse cells and their revertants. J. gen. Virol. *36:* 111–122 (1977).

71 Hughes, R.G.; Munyon, W.H.: Temperature-sensitive mutants of herpes simplex virus type 1 defective in lysis but not in transformation. J. Virol. *16:* 275–283 (1975).

72a Rapp, F.; Turner, N.; Schaffer, P.A.: Biochemical transformation by temperature-sensitive mutants of herpes simplex virus type 1. J. Virol *34:* 704–710 (1980).

72b Jamieson, A.T.; Macnab, J.C.M.; Perbal, B.; Clements, J.B.: Virus specified enzyme activity and RNA species in herpes simplex virus type 1 transformed mouse cells. J. gen. Virol. *32:* 493–508 (1976).

73 Wigler, M.; Sweet, R.; Sim, G.K.; Wold, B.; Pellicer, A.; Lacey, E.; Maniatis, T.; Silverstein, S.; Axel, R.: Transformation of mammalian cells with genes from procaryotes and eucaryotes. Cell *16:* 777–785 (1979).

74 Rapp, F.; Westmoreland, D.: Cell transformation by DNA-containing viruses. Biochim. biophys. Acta *458:* 167 (1976).

75 Flannery, V.L.; Courtney, R.J.; Schaffer, P.A.: Expression of an early, nonstructural antigen of herpes simplex virus in cells transformed *in vitro* by herpes simplex virus. J. Virol. *21:* 284–291 (1977).

76 Dreesman, G.R.; Burek, J.; Adam, E.; Kaufman, R.H.; Melnick, J.L.; Powell, K.L.; Purifoy, D.J.M.: Expression of herpesvirus-induced antigens in human cervical cancer. Nature, Lond. *283:* 591–593 (1980).

77 Bayliss, G.J.; Marsden, H.S.; Hay, J.: Herpes simplex virus proteins: DNA-binding proteins in infected cells and in the virus structure. Virology *68:* 124–134 (1975).

78 Wilcox, K.W.; Kohn, A.; Sklyanskaya, E.; Roizman, B.: Herpes simplex virus phosphoproteins. I. Phosphate cycles on and off some viral polypeptides and can alter their affinity for DNA. J. Virol. *33:* 167–182 (1980).

79 Aurelian, L.; Davis, H.J.; Julian, C.G.: Herpesvirus-type-2-induced tumor-specific antigens in cervical carcinoma. Am. J. Epidem. *98:* 1–9 (1973).

80 Aurelian, L.; Strnad, B.C.; Smith, M.F.; Immunodiagnostic potential of a virus-coded tumor-associated antigen (AG-4) in cervical cancer. Cancer, N.Y. *39:* 1834–1849 (1977).

81 Strnad, B.C.; Aurelian, L.: Proteins of herpesvirus type 2. II. Studies demonstrating a correlation between a tumor-associated antigen (AG-4) and a virion protein. Virology *73:* 244–258 (1976).

82 Gupta, P.; Rapp, F.: Identification of virion polypeptides in hamster cells transformed by HSV type 1. Proc. natn. Acad. Sci. USA *74:* 372–374 (1977).

83 Reed, C.L.; Cohen, G.H.; Rapp, F.: Detection of a virus-specific antigen on the surface of herpes simplex virus-transformed cells. J. Virol. *15:* 668–670 (1975).

84 Kimura, S.; Okazaki, K.; Yoshida, N.; Ohnishi, Y.: Effect of actinomycin D on the expression of herpes simplex virus-common surface antigen in cells transformed by herpes simplex virus type 2. J. Virol. *29:* 161–169 (1979).

85 Kutinova, L.; Vonka, V.; Broucek, J.: Increased oncogenicity and synthesis of herpes virus antigens in hamster cells exposed to herpes simplex virus type 2. J. natn. Cancer Inst. *50:* 759–766 (1973).

86 Boyd, A.L.; Orme, T.W.: Transformation of mouse cells after infection with UV-irradiation-inactivated HSV type 2. Int. J. Cancer *16:* 526–538 (1975).

87 Payne, G.S.; Courtneidge, S.A.; Crittenden, L.B.; Fadly, A.M.; Bishop, J.M.; Varmus, H.E.: Analysis of avian leukosis virus DNA and RNA in bursal tumors: viral gene expression is not required for maintenance of the tumor state. Cell *23:* 311–322 (1981).

88 Neel, B.G.; Hayward, W.S.; Robinson, H.L.; Fang, J.; Astrin, S.M.: Avian leukosis virus-induced tumors have common proviral integration sites and synthesize discrete new RNAs: oncogenesis by promoter insertion. Cell *23:* 323–334 (1981).

89 Frenkel, N.; Roizman, B.; Cassai, E.; Nahmias, A.: A DNA fragment of herpes simplex

and its transcription in human cervical cancer tissue. Proc. natn. Acad. Sci. USA *69:* 3784–3789 (1972).

90 McDougall, J.K.; Galloway, D.A.: Detection of viral nucleic acid sequences using in situ hybridization; in Stevens, Todaro, Fox, Persistent viruses. ICN-UCLA symposia on molecular and cellular biology, vol. XI, pp. 181–188 (Academic Press, New York 1978).

91 Kimura, S.; Esparza, J.; Benyesh-Melnick, M.; Schaffer, P.A.: Enhanced replication of temperature-sensitive mutants of herpes simplex virus type 2 at the nonpermissive temperature in cells transformed by HSV-2. Intervirology *3:* 162–169 (1974).

92 Macnab, J.C.M.; Timbury, M.: Complementation of ts mutants by a herpes simplex virus ts transformed cell line. Nature, Lond. *261:* 233–235 (1976).

93 Park, M.; Lonsdale, D.M.; Timbury, M.C.; Subak-Sharpe, J.H.; Macnab, J.C.M.: Genetic retrieval of viral genome sequences from herpes simplex virus-transformed cells. Nature, Lond. *285:* 412–415 (1980).

94 Hampar, B.; Ellison, S.A.: Chromosomal aberrations induced by an animal virus. Nature, Lond. *192:* 145–147 (1961).

95 Hampar, B.; Ellison, S.A.: Cellular alterations in the MCH line of Chinese hamster cells following infection with herpes simplex virus. Proc. natn. Acad. Sci. USA *49:* 474–480 (1963).

96 Stich, H.F.; Hsu, T.C.; Rapp, F.: Viruses and mammalian chromosomes. I. Localization of chromosomal aberrations after infection with herpes simplex virus. Virology *22:* 439–445 (1964).

97 Rapp, F.; Hsu, T.C.: Viruses and mammalian chromosomes. IV. Replication of herpes simplex virus in diploid Chinese hamster cells. Virology *25:* 401–411 (1965).

98 Waubke, R.; zur Hausen, H.; Henle, W.: Chromosomal and autoradiographic studies of cells infected with herpes simplex virus. J. Virol. *2:* 1047–1054 (1968).

99 O'Neill, F.J.; Rapp, F.: Early events required for induction of chromosomal abnormalities in human cells by herpes simplex virus. Virology *44:* 544–553 (1971).

100 Nachtigal, M.; Duff, R.; Rapp, F.: Chromosome aberrations in Syrian hamster embryo cells transformed after exposure to UV-irradiated herpes simplex virus type 1 and 2. J. natn. Cancer Inst. *54:* 92–105 (1975).

101 Mocarski, E.S.; Post, L.E.; Roizman, B.: Molecular engineering of the herpes simplex virus genome: insertion of a second L-S junction into the genome causes additional genome inversions. Cell *22:* 243–255 (1980).

102 Cairns, J.: The origin of human cancers. Nature, Lond. *289:* 353–357 (1981).

103 Yamanishi, K.; Ogino, T.; Takahashi, M.: Induction of cellular DNA synthesis by a temperature-sensitive mutant of herpes simplex virus type 2. Virology *67:* 450–462 (1975).

104 Melvin, P.; Kucera, L.S.: Induction of human cell DNA synthesis by herpes simplex virus type 2. J. Virol. *15:* 534–539 (1975).

105 Ejercito, P.M.; Kieff, E.D.; Roizman, B.: Characterization of herpes simplex virus strains differing in their effects on social behaviour of infected cells. J. gen. Virol. *2:* 357–364 (1968).

106 Hampar, B.; Aaronson, S.A.; Derge, J.G.; Chakrabarty, M.; Showalter, S.D.; Dunn, C.Y.: Activation of an endogenous mouse type C virus by ultraviolet-irradiated herpes simplex virus types 1 and 2. Proc. natn. Acad. Sci. USA *73:* 646–650 (1976).

107 Hampar, B.; Hatanaka, M.; Aulakh, G.; Derge, J.G.; Lee, L.; Showalter, S.: Type C virus activation in 'nontransformed' mouse cells by UV-irradiated herpes simplex virus. Virology *76:* 876–881 (1977).

108 Boyd, A.L.; Derge, J.G.; Hampar, B.: Activation of endogenous type C virus in BALB/C mouse cells by herpesvirus DNA. Proc. natn. Acad. Sci. USA *75:* 4558–4562 (1978).
109 Boyd, A.L.; Enquist, L.; Vande Woude, G.F.; Hampar, B.: Activation of mouse retrovirus by herpes simplex virus type 1 cloned DNA fragments. Virology *103:* 228–231 (1980).
110 Given, D.; Yee, D.; Griem, K.; Kieff, E.: DNA of Epstein-Barr virus. V. Direct repeats of the ends of Epstein-Barr virus DNA. J. Virol. *30:* 852–862 (1979).
111 Kieff, E.; Given, D.; Powell, A.L.T.; King, W.; Dambaugh, T.; Raab-Traub, N.: Epstein-Barr virus: structure of the viral DNA and analysis of viral RNA in infected cells. Biochim. biophys. Acta Rev. Cancer *560:* 355–373 (1979).
112 Kintner, C.R.; Sugden, B.: The structure of the termini of the DNA of Epstein-Barr virus. Cell *17:* 661–671 (1979).
113 Dambaugh, T.; Raab-Traub, N.; Heller, M.; Beisel, C.; Hummel, M.; Cheung, A.; Fennewald, S.; King, W.; Kieff, E.: Variations among isolates of Epstein-Barr virus. Ann. N.Y. Acad. Sci. *354:* 309–325 (1979).
114 Given, D.; Kieff, E.: DNA of Epstein-Barr virus. VI. Mapping of the internal tandem reiteration. J. Virol. *31:* 315–324 (1979).
115 Hayward, S.D.; Nogee, L.; Hayward, G.: Organization of repeated regions within the Epstein-Barr virus DNA molecule. J. Virol. *33:* 507–521 (1980).
116 Dambaugh, T.; Beisel, C.; Hummel, M.; King, W.; Fennewald, S.; Cheung, A.; Heller, M.; Raab-Traub, N.; Kieff, E.: Epstein-Barr virus (B95-8) DNA. VII. Molecular cloning and detailed mapping. Proc. natn. Acad. Sci. USA *77:* 2999–3003 (1980).
117 Skare, J.; Strominger, J.L.: Cloning and mapping of Bam HI endonuclease fragments of DNA from the transforming B95-8 strain of Epstein-Barr virus. Proc. natn. Acad. Sci. USA *77:* 3860–3864 (1980).
118 Miller, G.: Biology of Epstein-Barr virus; in Klein, Viral oncology, pp. 713–738 (Raven Press, New York 1980).
119 Adams, A.: Molecular biology of EBV; in Klein, Viral oncology, pp. 683–711 (Raven Press, New York 1980).
120 Moss, D.J.; Pope, J.H.: Assay of the infectivity of Epstein-Barr virus by transformation of human leucocytes *in vitro.* J. gen Virol. *17:* 233–236 (1972).
121 Henderson, E.; Miller, G.; Robinson, J.; Heston, L.: Efficiency of transformation of lymphocytes by Epstein-Barr virus. Virology *76:* 152–163 (1977).
122 Henderson, E.; Heston, L.; Grogan, E.; Miller, G.: Radiobiological inactivation of Epstein-Barr virus. J. Virol. *25:* 51–59 (1978).
123 Nonoyama, M.; Pagano, J.S.: Separation of Epstein-Barr virus DNA from large chromosomal DNA in non-virus-producing cells. Nature, Lond. *238:* 169–171 (1972).
124 Lindahl, T.; Adams, A.; Bjursall, G.; Bornkamm, G.W.; Kaschka-Dierich, C.; Jehn, U.: Covalently closed circular duplex DNA of Epstein-Barr virus in a human lymphoid cell line. J. molec. Biol. *102:* 511–530 (1976).
125 Adams, A.; Lindahl, T.; Klein, G.: Linear association between cellular DNA and Epstein-Barr virus DNA in a human lymphoblastoid cell line. Proc. natn. Acad. Sci. USA *20:* 2888–2892 (1973).
126 King, W.; Thomas-Powell, A.L.; Raab-Traub, N.; Hawke, M.; Kieff, E.: Epstein-Barr virus RNA. V. Viral RNA in a restringently infected, growth transformed cell line. J. Virol. *36:* 506–518 (1980).
127 Powell, A.L.T.; King, W.; Kieff, E.: Epstein-Barr virus-specific RNA. III. Mapping of DNA encoding viral RNA in restringent infection. J. Virol *29:* 261–274 (1979).

128 Delius, J.; Bornkamm, G.W.: Heterogeneity of Epstein-Barr virus. III. Comparison of a transforming and a nontransforming virus by partial denaturation mapping of their DNAs. J. Virol. *27:* 81–89 (1978).

129 Roizman, B.: Aspects of Epstein-Barr virus. Cell *19:* 562–563 (1980).

130 Schwartz, R.: Epstein-Barr virus – oncogen or mitogen? New Engl. J. Med. *302:* 1307–1308 (1980).

131 Albrecht, T.; Rapp, F.: Malignant transformation of hamster embryo fibroblasts following exposure to ultraviolet-irradiated human cytomegalovirus. Virology *55:* 53–61 (1973).

132 Geder, L.; Lausch, R.; O'Neill, F.; Rapp, F.: Oncogenic transformation of human embryo lung cells by human cytomegalovirus. Science *192:* 1134–1137 (1976).

133 Geder, L.; Rapp, F.: Evidence for nuclear antigens in cytomegalovirus-transformed human cells. Nature: *265:* 184–186 (1977).

134 O'Callaghan, D.J.; Henry, B.E.; Wharton, J.H.; Dauenhauer, S.A.; Vance, R.B.; Staczek, J.; Robinson, R.A.: Equine herpes viruses: biochemical studies on genome structure, DI particles, oncogenic transformation and persistent infection; in Becker, Herpes virus DNA (Nijhoff, The Hague, in press, 1981).

135 Robinson, R.A.; Henry, B.E.; Duff, R.G.; O'Callaghan, D.J.: Oncogenic transformation by equine herpesviruses. I. Properties of hamster embryo cells transformed by UV-irradiated EHV-1. Virology *101:* 335–362 (1980).

136 Robinson, R.A.; Vance, R.B.; O'Callaghan, D.J.: Oncogenic transformation by equine herpesviruses. II. Coestablishment of persistent infection and oncogenic transformation of hamster embryo cells by EHV-I preparation enriched for DI particles. J. Virol. *36:* 204–219 (1980).

137 Falk, L.A.: Biology of *Herpesvirus saimiri* and *Herpesvirus ateles;* in Klein, Viral oncology, pp. 813–832 (Raven Press, New York 1980).

138 Falk, L.A.; Wright, J.; Wolfe, L.G.; Deinhardt, F.: *Herpesvirus ateles:* transformation *in vitro* of marmoset spleen lymphocytes. Int. J. Cancer *14:* 244–251 (1974).

139 Fleckenstein, B.; Mulder, C.: Molecular biological aspects of *Herpesvirus saimiri* and *Herpesvirus ateles;* in Klein, Viral oncology, pp. 799–812 (Raven Press, New York 1980).

Dr. David M. Knipe, Department of Microbiology and Molecular Genetics, Harvard Medical School, 25 Shattuck Street, Boston, MA 02115 (USA)

Prog. med. Virol., vol. 28, pp. 145–179 (Karger, Basel 1982)

Epstein-Barr-Virus Nuclear Antigen

Vladimír Vonka, Ivan Hirsch

Department of Experimental Virology, Institute of Sera and Vaccines, Prague, Czechoslovakia

Contents

I. Introduction

Epstein-Barr (EB) virus has been identified as the etiological agent in infectious mononucleosis (IM) and has been implicated in the pathogenesis of at least two human malignancies, Burkitt lymphoma (BL) and nasopharyngeal carcinoma (NPC). The evidence for its association with the latter two tumors is based on the elevation of immune responses to EB viral antigens in the patients, a transformation potential in EB virus for B lymphocytes of human or simian origin, regular findings of EB viral fingerprints in the tumor cells, an oncogenic potential of EB virus for some New World monkeys, as well as outcome of a recent prospective study on BL.

All cells carrying the EB virus genome contain the nuclear antigen of EB virus (EBNA). This substance, also denoted soluble antigen (SA) of EB virus, is at present in the focus of interest. Because of its many similarities with the T antigens of papova- and adenoviruses, EBNA is suspected of playing an essential role in the processes of cell transformation by EB virus and maintenance of the transformed state.

In this review we shall survey the past studies on this substance as well as the present knowledge about its properties and possible biological role in the infected cell.

II. Early Studies: Discovery of the Soluble Antigen (SA) of EB Virus

The first report on the presence of a complement-fixing (CF) SA in lymphoblastoid cell lines was published by *Armstrong* et al. [4] in 1966; however, its relationship to EB virus was not recognized at that time. 3 years later, the SA present in the high-speed centrifugation supernatants of extracts from various lymphoblastoid cell lines was clearly shown to be associated with EB virus independently in three laboratories and it was suggested that it is an EB-virus-coded, but non-structural, antigen [39, 114, 165–167]. This conclusion was based on the following observations:

(i) The majority of sera possessing antibody to the viral capsid antigen (VCA) of EB virus, as determined by indirect immunofluorescence (IF) [48], were reactive with SA in the CF test; all sera negative in the IF test were also non-reactive in the CF test. This indicated that the presence of SA antibody is conditioned by previous infection with EB virus.

(ii) SA-antibody-positive sera did not react with antigens similarly prepared from human non-lymphoblastoid cell lines.

(iii) The antibodies reactive with VCA or SA were clearly different. Although most of the sera containing high SA antibody titers also possessed high levels of VCA antibody, many sera with high VCA antibody levels were devoid of detectable SA antibody. On the other hand, the reactivity of the same sera with a CF antigen consisting of purified viral nucleocapsids strongly resembled that seen in the IF test. Moreover, the reactivity of human sera with VCA in the IF test could not be blocked by excess of SA.

(iv) There was no quantitative relationship between the virus particle count (and the percentage of IF-VCA-positive cells) and the SA content in various cell lines tested, and SA could be separated from EB-virus particles by centrifugation of cell extracts. In addition, the antigen could also be detected in lymphoblastoid cell lines which do not produce the virus, such as Raji or NC37. Later on, these cells were shown to possess EB virus DNA [103, 179].

(v) SA also seemed different from the membrane antigen (MA), as defined at that time, since the latter was only detected in cell lines producing EB virus [71] and since it was impossible to absorb SA antibody by MA-positive cells.

Early attempts to characterize SA [39, 166, 172] revealed the following:

(i) SA is thermostable; a considerable amount of activity was preserved even after prolonged storage at 37 °C and after short-time heating up to 80 °C.

(ii) The antigen is stable over a wide pH range.

(iii) It is highly trypsin-sensitive.

(iv) SA production seemed dependent on the cell growth rate; under conditions restricting cell proliferation its content decreased.

Nearly simultaneously with the recognition of SA by the CF test, several papers appeared reporting on the presence in cell extracts of antigens detectable by immunodiffusion [19, 104, 124, 145]. The heat-resistant antigen found in both producer and non-producer cells appeared to be most probably identical with SA.

III. Discovery of EB-Virus Nuclear Antigen (EBNA) and Evidence for Its Identity with SA

Attempts to localize SA in cells of lymphoblastoid cell lines by the direct or indirect IF test failed. Early efforts to amplify the antigen-antibody reaction by complement and to visualize it by means of anti-complement immunofluorescence (ACIF) were also unsuccessful [34], probably owing to inadequate quality of the reagents used. Later on, *Reedman and Klein* [122] using the same approach reported the presence of an intranuclear antigen, which they denoted EBNA. This antigen was bound to metaphase chromosomes. In interphase nuclei it was demonstrated on chromatin fibrils by means of peroxidase-labelled antibodies and electron microscopy [6, 135] and later was found to be associated with purified chromatin isolated from EBNA-positive cells [112]. EBNA was gradually shown to be present in the cells of all lines positive for EB-virus DNA [122], in BL biopsies [85, 123], in epithelial cells of NPC [59, 68], in biopsies from carcinoma of the nasal fossa [60] in simian tumors induced by EB virus [77, 99], in up to 2% of the B-cell fraction of the large blast cells present in the peripheral blood of IM patients [72], in lymph node cells of a child with severe primary EB virus infection [82], in a significant proportion of B cells present in the tonsils of subjects suffering from chronic exudative tonsillitis [161, 162], in B-type cells of primate origin infected with EB virus in vitro [5, 27, 96], in somatic cell hybrids between EB-virus-genome-carrying cells and EB-virus-negative cells of human, rodent or simian origin [44, 45, 144, 159], in human amnion epithelial cells microinjected with EB virus concentrates [70], and also in various types of EB-virus-infected human, murine and baboon cells onto which EB-virus receptors had been transplanted prior to the infection [163]. On the other hand, EBNA was regularly absent from cells of normal tissues, biopsies from EB-virus-genome-negative tumors and cell lines derived from these tissues, and absent also from EB-virus-genome-negative lymphoblastoid cell lines. Nearly all subjects who had experienced EB-virus infection, as indicated by the presence of VCA antibody, possessed EBNA antibody, although frequently in serum dilutions less than 1:10. EBNA antibody was absent from subjects free of VCA antibody [122]. All these findings strongly suggested that EBNA is an EB-virus-specific product.

A considerable body of evidence exposing the identity of EBNA and SA has been obtained. The first indication of this was the demonstration of a very high degree of concordance of anti-EBNA titers with anti-SA titers [73]. Further findings supporting the close association of EBNA with SA were obtained by *Lenoir* et al. [80]. They characterized SA in terms of sucrose-gradient centrifugation, gel-filtration and ion-exchange chromatography. All fractions which contained SA activity, as determined in the CF test, inhibited the ACIF reaction used to detect EBNA. Using isoelectric focusing of SA and chromatin-associated EBNA, *Pikler* et al. [112] revealed a single peak for both activities at pH 4.6. The specific absorption of EBNA antibody by SA antigen was reported by *Luka* et al. [89]. Further tests indicated that the EBNA reaction could be reconstituted by adding SA to EBNA-negative nuclei [109] and that in isolated nuclei the EBNA reaction was abolished under similar conditions, as when SA was eluted from DNA-cellulose columns [56].

On the basis of these results it is almost generally accepted that EBNA and SA, or at least their major component, are a single substance. The rare discrepancies occasionally reported [22, 84, 142] were most probably due to the incapability of the CF test to detect low levels of antibody, or to the interference of anti-nuclear factors, or to the use of extracts from productive cell lines possibly containing, in addition to EBNA, other EB-virus-associated antigens. Apart from EBNA also VCA [38, 166], the early antigen (EA) [79] and possibly other EB viral antigens can function as CF antigens.

The antigen under discussion will accordingly henceforth be referred to as EBNA in this paper. The term SA will only be used when reference is made to CF tests using cell-extract antigens. It should be added that apart from ACIF on fixed cells and CF tests with cell extracts, other techniques have also been developed for monitoring EBNA and EBNA antibody. These include an electron microscopy test utilizing peroxidase-labelled anti-complement antibody [135], a ^{125}I-IgG binding assay [12, 13], a solid-phase radioimmunoassay [25], ELISA test [133, 171] and a single radial immunodiffusion test [93]. Although some of these tests represent rapid and highly sensitive assays, they have not been extensively utilized in the studies on EBNA. Apart from serological techniques, reactivity to EBNA can also be revealed by the leukocyte migration inhibition test [150].

IV. Significance of the Presence of EBNA Antibody

As already mentioned, nearly all those who have been infected with EB virus possess EBNA antibody. Once developed, it tends to persist through life. It can be assumed that the continuous processes of proliferation of EBNA-positive cells and their destruction by immunological means provide the antigen for permanent stimulation. The relative stability of EBNA antibody titers suggests that these two processes are well balanced.

For the past 12 years the EBNA antibody status has been intensively investigated in healthy subjects as well as in those suffering from a variety of diseases. It was also determined in various ethnic groups; these studies were primarily aimed at finding relationship between the epidemiological characteristics of EB virus and the distribution of the different EB virus-associated diseases.

A. Delayed Appearance of EBNA Antibody in Primary EB Virus Infection

The first studies on the presence of antibody to EBNA were performed by CF test and predominantly concerned with sera from healthy subjects [114, 166]. The data obtained indicated that a majority of those who had experienced EB virus infection, as monitored by their reactivity in the indirect IF test, possessed SA antibody. However, when different age groups were followed one notable exception was observed: nearly all subjects aged 6–12 months who possessed VCA antibody were free of SA antibody. Since these infants represented subjects who most probably had been infected with EB virus only recently, we suggested that SA antibody appears considerably later in the course of infection than VCA antibody does [166].

To explore this point further, we examined sera from patients suffering from IM, i.e., a clinical manifestation of primary infection by EB virus [see the recent review, 50]. Already the first results indicated that SA antibody was absent from sera taken from patients up to several months after the onset of the disease [164]. This observation was confirmed and corroborated by subsequent investigations [139, 140, 147, 168]. In the study carried out in our laboratory [168], the presence of VCA and SA antibodies in 159 sera from IM patients were examined. All but one serum taken from 1 patient on the 1st day of disease were positive for VCA antibody. On the other hand, only 1 of 97 sera taken within the 1st month and only 1 of 7 sera taken 4–5 months after onset of disease possessed SA antibody. Thus the SA antibody patterns in IM patients in the first months following the onset of disease strongly resembled those observed in VCA-antibody-positive children aged 6–12 months. Only after periods exceeding 6 months from the onset of disease was SA antibody detected as frequently as in normal subjects older than 1 year.

These data obtained in the 'pre-EBNA' period were subsequently confirmed by others using the ACIF test [24, 42, 51, 142]. *Henle* et al. [51] reported that only after periods exceeding 6 months did all patients possess EBNA antibody; still they were at a lower level by that time than the antibody seen in normal blood donors. The concordance between the CF-SA and ACIF-EBNA tests in IM patients is a further piece of evidence in favor of the identity of SA and EBNA.

Some more evidence on the delayed development of SA antibody was obtained by *Holý* et al. [58], who followed the spread of herpesviruses in a closed community of children. In this study serial blood samples were examined for viral antibodies. Only exceptionally was SA antibody detected prior to the 5th month after seroconversion to a VCA-positive state. Similar data were recently reported by *Biggar* et al. [11] in African children. In most of them EBNA antibody was found not earlier than 5 months after seroconversion to VCA-positivity.

The reasons for the late development of EBNA antibody in EB virus infection are not conclusively understood. Several explanations may be offered:

(i) EBNA is a weak antigen, and long-lasting stimulation may be required for antibody development. The difficulties associated with the preparation of EBNA antibody in animals [unpublished data] are in line with this conjecture.

(ii) Not enough antigen is available in the early phases of infection. This would certainly not be due to the absence of EBNA-positive cells, because up to 2% of cells present in the T-cell-depleted peripheral lymphocyte population of acute IM patients are EBNA-positive [72]. However, at this stage the antigen may be present in a form which is not sufficiently immunogenic. Although considerably thermostable in cell extracts, EBNA seems to be labile within the cells [33, 110, 148, 165], and little antigen may be available when the cell succumbs to productive infection. It may take a rather long time before EBNA-positive cell clones start to proliferate.

(iii) Although specifically sensitized killer T lymphocytes are present in the acute stage of IM [149], the immunological mechanism operative in the recognition and destruction of the EBNA-positive transformed cells in vivo may also be a delayed one. It should be recalled that it takes up to several months after onset of disease before the immunological functions are fully operative [130]. The mechanisms involved in the control of EBNA-positive transformed cells and release of sufficient amounts of the antigen are largely unknown; however, they are most probably associated with a T-cell function. The recent observation that ataxia teleangiectasia patients, known to suffer from a T-cell deficiency, frequently fail to develop EBNA antibody in spite of markedly elevated VCA and EA antibody titers [10], is consistent with this assumption.

(iv) EBNA antibody may be formed early after infection but is absorbed by an antigen excess released from the infected cells. This does not seem likely because EBNA antibody is missing in IM patients a long time after the disappearance of EBNA-positive cells from the blood stream.

Regardless of which explanation ultimately proves correct, testing sera for EBNA antibody is helpful in differential diagnosis where IM is suspected. Its presence seems to be incompatible with the diagnosis of IM caused by EB virus.

B. EBNA Antibody in Patients with Various Malignant Diseases

In patients suffering from malignant diseases in which EB-virus etiology has been implicated, the EBNA antibody status is markedly different from that seen in IM patients. The first systematic study on the evolution of SA antibody in BL patients was reported by *Sohier and de Thé* [141]. According to their results, sera from patients whose tumors progressed in the course of the observation period exhibited high and/or increasing titers, while the sera of those in a long-term remission possessed low and/or decreasing SA-antibody titers. *Nkrumah* et al. [102] determined antibodies to both structural and non-structural antigens of EB virus in serial samples of 141 BL patients. When comparing the survivors with those who had died in the course of the observation period, they found significantly higher titers against both EA and EBNA in the latter group of patients. They also detected higher EBNA antibody titers in patients with positive skin reactions to extract from Raji cells (an EB-virus-genome-positive and EBNA-positive, non-producer BL line) than in those who lacked this reactivity.

Data which suggested an association between the tumor progression and the EBNA-antibody status, appear to be at variance with the findings in subjects who developed BL in the course of the prospective study on this disease sponsored by the International Agency for Research on Cancer (IARC): no marked differences were determined between their pre-illness and illness sera [21]. However, this may have been associated with the fact that these patients were probably bled early after the clinical manifestation of the disease. The search for new BL cases in this study was very intensive; they were actively looked for by a team of specialists who regularly visited all health centers. It is therefore likely that the enhanced EBNA-antibody levels observed in earlier studies reflected a heavy tumor burden rather than only the progression of the disease.

Several reports are available on the EBNA antibody status in NPC patients. The first to appear was concerned with a group of 15 Chinese NPC patients and appropriate controls [164]. Both VCA and SA antibodies were markedly higher in the patients than in control groups comprising either normal individuals or patients suffering from a different cancer. *De Thé* et al. [23] followed the development of SA antibody in 37 NPC patients. The antibody titers were markedly elevated not only over those found in normal control subjects but also over those determined by the same investigators in BL patients [141]. At variance with the findings in the latter group of patients, no relationship between antibody level and progression of disease was apparent, although regression of disease was in some patients associated with a decrease in antibody titers. When investigating another group of 38 NPC patients, the same authors [22] were capable of demonstrating a dependence between progression of the disease and the development of EBNA and SA antibodies, although their rises were less steep than in the case of EA antibody. Also *Huang* et al. [60] found higher EBNA-antibody levels in NPC patients than in those suffering from other tumors of the head and neck.

The most extensive study on the EB virus antibody status of NPC patients was carried out by *Henle* et al. [52]. They followed 104 Chinese patients for periods exceeding 4 years with the aim to elucidate the relationship between the levels of different antibodies and the clinical stage of the disease. While there was a clear relationship between an increase in VCA and EA antibodies and the progress of the disease (stages I + II versus stages III + IV), this was not true for the EBNA antibody. In fact, its geometric mean titer (GMT) in the patients with the most advanced stage of the disease (stage IV) was somewhat lower than in patients with less developed disease (stages I–III). In addition, in a number of patients in whom the disease had progressed to death and in whom EA and VCA antibodies had been gradually increasing, or in patients who had responded well to therapy and in whom the EA- and VCA-antibody titers had decreased, the EBNA-antibody levels remained relatively stable throughout the follow-up.

The EBNA antibody status was also investigated in some non-Chinese NPC patients [84]. All patient groups irrespective of race and country of origin possessed not only increased levels of VCA and EA antibodies but also somewhat higher EBNA antibody titers than patients suffering from other cancers. However, the titers in Chinese NPC patients, included as controls, were higher than those detected in Tunisian, French or German patients.

Antibodies to different EB virus antigens have been followed in a group of Czechoslovak NPC patients for more than 9 years now [*Suchánková* et al., to be published]. Generally, a pattern similar to that reported by *Henle* et al. [52] has been observed. However, a greater variation in SA-antibody titers in individual patients has been recorded throughout the observation period.

Several studies indicated raised levels of VCA antibody [63, 83] and also EA antibody [49, 54, 131] in Hodgkin lymphoma (HL) patients. These observations are usually interpreted as a consequence of an impaired cell-mediated immunity associated with the disease and with the use of immunosuppressive drugs, permitting more extensive EB virus replication. Thus they do not seem to have any etiological significance. *Rocchi* et al. [132] found that anti-EBNA reactivity was also higher in HL patients than in controls, more so in those suffering from the lymphocyte-depletion condition than from the other forms of HL. Since the lymphocyte-depletion subtype of the disease is usually associated with low-grade cell immunity [175], this finding seems to be in accord with the above-mentioned concept. It is not in contradiction with the low levels or absence of EBNA antibody in patients suffering from inherent immunodeficiencies such as ataxia teleangiectasia or Behçet disease [10], since in those with acquired immunodeficiencies the primary response to the EB virus antigens was most likely not altered.

EBNA antibody was also determined in non-Hodgkin, non-Burkitt lymphoma patients [26]. At variance with the above-mentioned groups of patients, no difference between patients and controls was observed, in spite of moderate differences in VCA and EA antibodies.

Tonsillar carcinoma patients, in whom both VCA and EA antibodies were found more frequently and at higher titers than in matched controls, displayed raised EBNA-antibody levels as well [169]. The presence of EB-virus DNA detected in 4 out of 9 biopsy specimens was associated with somewhat elevated EBNA antibody in the respective patients [*Brichácek* et al., to be published]. Increased levels of antibodies to EB-virus antigens, including EBNA, have also been reported in patients suffering from chronic exudative tonsillitis associated with the presence of EBNA-positive cells in the tonsillar tissues [161].

C. EBNA Antibody in Various Ethnic Groups

In a study executed in collaboration with the IARC, we examined SA-antibody profiles in various populations in Europe, Asia and Africa [20]. In the population of Uganda, with early and massive EB-virus infections, the majority of children aged 1–3 years already possessed SA antibody. In the other populations studied, in which the EB-virus infections were delayed, the SA antibody started to appear at a later age. The most interesting single observation related to EBNA was concerned with differences in the prevalence of SA antibody in the Chinese and the non-Chinese populations living in the same locality. While the Indians, living in Singapore, tended to experience the EB-virus infection earlier and possessed higher VCA-antibody titers than their Chinese fellow countrymen, the latter group possessed SA antibody more frequently and in higher titers than the former one. It can be speculated that this differential antibody response to EBNA is associated, in a way, with the different course and outcome of EB-virus infection in these ethnic groups. NPC is much more common among the Chinese than among the Indians. It is possible, e.g., that EBNA-positive cells are more frequent in the Chinese than the Indians because of a genetically conditioned less strict immunological control over these cells or because of genetically determined differences in EB virus-cell interactions. It would be of interest to compare the tranformability by EBvirus of lymphocytes from the Chinese and from the Indians (or from other races with comparably low incidence of EB virus-associated malignant diseases). To our knowledge such experiments have not yet been carried out.

V. Relationship between EBNA and Nuclear Antigens of Other Herpesviruses

Nuclear antigens similar to EBNA have also been detected in cells transformed by other herpesviruses. Among the oncogenic animal herpes viruses, Marek's disease virus is an exception: in cells transformed by this virus an EBNA-like antigen has not been demonstrated. However, this does not need to signify its absence but may be due to an incapability of the present techniques to detect it.

EB virus exhibits a high degree of genetic homology with the B lymphotropic herpesviruses of the Old World monkeys [2, 32, 41, 106, 120]. It is, therefore, not surprising that both their structural and non-structural antigens are cross-reactive. EBNA is antigenically related to the nuclear antigens of the herpesvirus of chimpanzees *(H. pan)* [40, 41] and that of baboons *(H. papio)* [106, 107]. This relationship appears to be a non-reciprocal one: EBNA antibody reacts with the nuclear antigens of chimpanzee and baboon herpesviruses both in the CF and the ACIF test, but sera of these animals do not react with EBNA. Absorption of EBNA antibody with EBNA abolished the reactivity with nuclear antigen of both EB virus and *H. papio,* but absorption with the nuclear antigen of *H. papio* removed the reactivity with this antigen but did not significantly reduce reactivity with EBNA. On the other hand, the nuclear antigens of the chimpanzee and baboon viruses are mutually closely related.

EBNA is also related to the nuclear antigens of the herpesviruses isolated from orangutans *(H. pongo)* and gorillas *(H. gorilla)* [2, 120]. A significant proportion of human sera reacted with the nuclear antigens of both viruses. None of the gorilla sera gave a positive reaction with EBNA, while one-third of orangutan sera were reactive with this antigen. The absorption of human sera with *H. gorilla*-nuclear antigen removed antibodies to the nuclear antigens of both homologous and heterologous simian viruses but not EBNA antibody [120]. These results confirm the asymmetrical relationship between EBNA and the nuclear antigens of various simian herpesviruses, and suggest that EBNA possesses at least two determinants, one of which is EBNA-specific while the other is shared with other nuclear antigens

The T-lymphotropic herpesviruses of the New World monkeys, *H. ateles* and *H. saimiri,* which have common DNA sequences and cross-reactive structural components, induce nuclear antigens with closely related antigenic determinants [2, 108]. No cross-reactivity with EBNA has been observed.

VI. Biochemical Characterization of EBNA

A. Purification of EBNA

Early attempts to purify EBNA gave only limited success, without direct identification of the partially purified protein with EBNA. In these experiments *Becker* et al. [9] extracted EBNA with 1.0 *M* NaCl or 0.03 *M* sodium deoxycholate, *Twardzik* et al. [160] introduced fractionation on hydroxylapatite, and *Ng* et al. [101] used immunoaffinity chromatography. At that time the most efficient and straightforward approach was employed by *Baron* et al. [7]. These authors used ammonium sulfate precipitation, DEAE cellulose and poly C-cellulose-column chromatography and achieved approximately 100-fold purification of EBNA.

It was not until recently that EBNA has been purified to biochemical homogeneity and partially characterized. The extraction methods used by most investigators resemble those introduced for SA extraction, with two notable modifications, namely the use of *(i)* high ionic strength [88, 89, 92] and *(ii)* higher pH [1, 86]. These conditions have been employed to facilitate the release of EBNA from EBNA-DNA complexes. In most purification procedures advantage has been taken of the exceptional thermostability [8, 86, 92] and DNA-binding property of EBNA [7, 8, 86, 88, 89, 92]. In addition to

DNA-cellulose, several other systems including hydroxylapatite [86, 88, 160], DEAE-Sephadex [80, 81, 86], blue dextran Sepharose [88], molecular sieving [80, 81, 88], immunoaffinity purification [92] and precipitation with ammonium sulfate [8, 92, 173] have been employed. In experiments representing probably the most successful arrangement of separation procedures, EBNA was purified 1200-fold, with a 45% yield of the original antigen amount [88]. At the same time other biophysical, chemical and immunological techniques have been utilized for EBNA characterization.

These studies have gradually provided the basic data on the physicochemical nature of EBNA. As estimated from hydrodynamic data and gel-filtration analysis, its native molecular weight is 170–230 kilodaltons (170K to 230K) [80, 81, 93, 109]. In a sucrose gradient containing 0.15 *M* NaCl, it sediments as a single peak with a sedimentation coefficient of 8.5S [80, 81]. Isoelectric focusing revealed that EBNA is an acid protein with an isoelectric point at pH 4.6 [93] or 4.8 [112]. Since it dissociates in buffers containing 0.5–1.0 *M* NaCl into components with molecular weights of about 100K, *Ohno* et al. [109] proposed that EBNA is composed of two smaller subunits. The subsequent studies have suggested a tetrameric structure for EBNA: SDS-PAGE of immunoprecipitates obtained by treating extracts of radioactively labelled EB-virus-genome-positive cells with EBNA-antibody yielded two molecular bands with molecular weights of 48K and 53K, respectively [70, 86]. Originally, the 48K polypeptide was considered to be a degradation product of the bigger one [70]. However, later results ruled out this possibility. *Luka* et al. [86] succeeded in demonstrating that only the 48K subunit carries the immunological specificity of EBNA. Moreover, although some similarities in the chemical composition of the 48K and 53K polypeptides (namely an excess of hydrophilic residues and a low content of hydrophobic and basic residues) were revealed, amino acid analysis and peptide mapping clearly differentiated between them [86]. Since the 48K and 53K proteins coprecipitate with EBNA-antibody, at least a portion of the native tetrameric EBNA complexes is most probably constituted by both of them. Separated proteins consisting of either 48K or 53K subunits are probably also present in tetrameric structures with molecular weights of about 200K [86]. Even denatured 48K subunits isolated after SDS-PAGE are able to regain their tetrameric structure and antigenic activity [53]. It can preliminarily be concluded that the 53K component is a cell polypeptide resembling proteins of similar size found in SV40-transformed cells (in complex with the large T antigen) and also in some other cells transformed by viruses or chemical carcinogens [18, 76].

A different approach for the characterization of EBNA was adopted by *Pikler* et al. [112, 113] and *Gilbert* et al. [43]. Based on the observation that only a portion (about 70%) of EBNA is eluted from the nuclei by 0.5 *M* NaCl, while more drastic treatment is needed for releasing the rest of it, they discriminate between 'soluble' EBNA (or EBNA I) and DNA-tightly-bound EBNA (which they denote EBNA II). Both possessed the immunological specificity of EBNA as determined by the indirect single radial immunodiffusion technique, and for each of them three major molecular species have been detected [112]. In subsequent tests, in which EBNA II was extracted with 3 *M* guanidine hydrochloride, these investigators showed that it shared most of its biochemical properties with histones. They even claim to have succeeded in absorbing EBNA antibody from human sera by isolated histones [43; *Pearson and Spelsberg*, personal commun.]. These interesting results require confirmation.

A certain variation, concerning not only quantitative but also qualitative aspects of EBNA purification, is observed when the results from different laboratiories or results of repeated experiments in the same laboratory are compared. Thus, apart from 48K and 53K subunits, additional components were occasionally found in purified EBNA preparations [92]; either one or both of the 48K and 53K bands were observed in SDS-PAGE of the immunoprecipitates formed with the same EBNA-antibody-positive serum [86, 88]; a remarkable variation in the phosphate concentration needed for EBNA elution from hydroxylapatite was also reported [86, 88, 160].

What is the source of this variation? Several factors in the course of the extraction and purification procedures may be of particular importance for the outcome of the experiments and thus for the reproducibility of the results.

(i) Depending on the extraction procedure used, crude EBNA preparations contain different amounts of contaminating DNA-binding proteins and cellular DNA. Because of the competitive character of the EBNA-DNA interactions [57], EBNA binds only poorly to DNA-cellulose in an excess of DNA in the solution. Moreover, data have been obtained suggesting that DNA may induce conformational changes in the EBNA molecule [8]. It is nearly certain that the chromatographic characteristics of EBNA-DNA complexes differ from those of EBNA itself.

(ii) EBNA monomers can apparently aggregate with different protein molecules as well as among themselves; the conditions of these reactions are largely unknown. Thus unequal rates of aggregation occurring at different purification steps may also substantially influence the outcome of the assays.

(iii) EBNA content within cells is a function of their physiological state

[29, 138, 165]. In the non-growing cultures it rapidly disappears, due to its apparent lability in the cell in contradistinction to its extreme thermostability in cell extracts. The degradation products of EBNA may participate in the above-mentioned interactions and thus interfere with their outcome. It can be assumed that not only the content but also the physicochemical nature of EBNA may be influenced by the state of culture from which it has been extracted.

A better understanding of the biochemical nature of EBNA and its interactions with other cell components is therefore needed before the purification procedures can be standardized.

B. Interaction of EBNA with DNA

Enzymatic or other biochemical activity of EBNA is not known at this writing. The most salient among all the biochemical features of EBNA described up to now is its DNA-binding character. EBNA binds tightly to double-stranded DNA [7, 8, 89], but only weakly to single-stranded DNA [89]. Low ionic strength and pH 6.0 have been found to be the optimal conditions for EBNA binding to DNA [56, 89]. EBNA has been eluted from double-stranded calf-thymus DNA-cellulose column as one homogenous, symmetrical peak with 0.38 *M* NaCl [89]. It failed to bind to DNA above pH 8.0 [56].

1. EBNA Binding to Acid-Fixed-Nuclei (AFNB)

Owing to its DNA-binding property, extracted EBNA can be bound to fixed nuclei of any species [109]. This reaction is usually called acid-fixed nuclear binding (AFNB). The fixation procedure is essential for the removal of the resident-DNA-associated proteins from the chromatin, as they would otherwise block the binding of exogenous EBNA. AFNB represents an amplification of the common ACIF test, most probably thanks to the ability of DNA in fixed nuclei to bind higher amounts of EBNA than are normally present in EB-virus-genome-positive nuclei. This has facilitated the detection of EBNA-like antigens also in cells transformed by simian herpesviruses, such as *H. papio* and *H. ateles* [105–108] and in tumor biopsies [87].

2. Problems Involved in Establishing Specificity of the Reaction

Of the various aspects of EBNA-DNA interaction, the question of specificity of this reaction seems to be of the greatest importance. The hypothesis that at least some biological effects of EBNA stem from its binding to regulatory DNA sequences assumes a certain degree of specific interaction between EBNA and these sequences, as has been found for the SV40T antigen [62, 121,125, 156, 157]. The first observations did not provide any evidence for the specificity of the DNA-EBNA interactions: EBNA was found equally distributed among all metaphase chromosomes [122]; this even applied to the AW-Ramos cells carrying only one EB-virus-genome copy per cell [3]. In cell hybrids between EB-virus-positive human and EB-virus-negative mouse cells, EBNA was likewise found associated with both the human and the mouse chromosomes [74, 144].

However, these results were not totally unexpected. It may be useful to recall the earlier findings from some other systems, in which specific DNA-protein interactions were demonstrated. These studies revealed two characteristic features of these interactions. First, the affinity of regulatory proteins for their target DNA sequences is usually very high. For example, the ratio of the specific to nonspecific association constant for the *lac* repressor on *E. coli* DNA is of the order of 10^6 [37]. Reconstitution studies with the T antigen of the ND2 hybrid virus [156] suggest an average affinity constant for SV40DNA greater than 10^8 M. Second, only few molecules are usually involved in the interactions with specific DNA sequences [118, 156]. Thus, the specific reaction can easily be obscured by the nonspecific interactions, which are much more numerous. The development of the knowledge of the interactions of SV40T antigen with DNA can again serve as an example: the nitrocellulose binding assay failed to distinguish between the binding of SV40T antigen to SV40DNA, ΦX174 RFI DNA, lambda DNA and plasmid pMB9 DNA [61, 156], although other methods finally demonstrated specific binding to SV40DNA beyond all doubt [62, 125, 156, 157].

Having in mind these problems we studied the affinity of EBNA for various DNAs by blocking the AFNB reaction by different dilutions of DNAs [55]. The underlying assumption was that with the use of highly diluted EBNA against relatively large amounts of DNA the predominant binding to specific versus nonspecific DNA sequences might be expressed. The AFNB method was employed because it permitted detection of specific reaction in spite of semi-purified EBNA being used. EBV DNA was found to be about twice as efficient as herpes simplex virus DNA and about three times more efficient than various cellular DNAs in abolishing the ability of DNA-cellulose-purified EBNA to convert acid-fixed nuclei to the EBNA-positive form.

In spite of having demonstrated some differences in EBNA affinity for different DNAs, we still do not consider them to be convincing enough to substantiate the conclusion that EBNA binds specifically to EB-virus DNA. To elucidate this point other methods, better capable of discriminating between specific and nonspecific bindings, have to be used. Such experiments are under way in our laboratory as well as in some other laboratories.

3. Interactions of EBNA-Anti-EBNA-Antibody Complexes with DNA and EBNA-DNA Complexes with Anti-EBNA Antibody

Crude preparations of EBNA always contain some cellular DNA bound to it. Because methods monitoring the presence of EBNA are based on its reaction with specific antibody, it would be of particular interest to explore the effect of DNA-binding properties of EBNA. Using AFNB, we examined whether EBNA contains distinct sites for interaction with antibody and with DNA. This investigation revealed the following [57]:

(i) While EBNA antibody reacts with EBNA already associated with DNA (i.e., the usual AFNB reaction), EBNA-anti-EBNA complexes would not bind to DNA.

(ii) While EBNA was readily eluted from acid-fixed nuclei by 0.4 *M* NaCl, it was not eluted even by 1.5 *M* NaCl if it had been exposed to EBNA antibody.

These results indicate that either the same sites of the EBNA molecule are responsible for both the antigenicity and the DNA-binding property, or that the reaction with antibody induces conformational changes in the EBNA molecule which substantially modify its reactivity with DNA. It should be recalled that DNA binding to EBNA influences the reactivity of EBNA with antibody as well. It was shown by *Baron and Strominger* [8] that about 50% of the complement-fixing activity was lost after treatment of an EBNA preparation with different DNases. The activity lost after DNase treatment could be restored, however, by the addition of calf-thymus DNA. This is another indication that DNA can induce conformational changes in the EBNA molecule resulting in its changed antigenicity: whereas antibody treatment decreases EBNA reactivity with DNA, interaction of EBNA with DNA seems to enhance its antigenicity.

VII. Biology of EBNA

A. Origin of EBNA

The critical experiments, such as the transcription and translation of polypeptides displaying EBNA antigenic properties from EB viral DNA in a cell-free system, which would directly prove the viral origin of EBNA, have not yet been performed. Similar experiments, for example, demonstrated beyond all doubts the viral origin of SV40 T antigen [117]. However, several lines of indirect evidence, as shown in the previous sections of this paper, strongly suggest that EBNA is specified by viral DNA. If this is true, then the corresponding genetic information should be localized within those EB-virus-DNA segments (of approximately 5×10^6 daltons) transcribed into the mRNA of non-producer lines [111, 115]. Although the efforts to map these sequences have provided somewhat conflicting results, the bulk of the present evidence indicates that they are encoded almost entirely in the Eco RI fragments A and B [66, 115, 116, 134]. Even though the coding capacity of the DNA sequences transcribed into the mRNA exceeds five medium-sized polypeptides, apart from the poorly defined lymphocyte-detected membrane antigen (or LYDMA) [149], only EBNA has been identified in non-producer cells thus far.

B. EBNA in Primarily Infected Cells

EBNA is present in cells of all EB-virus-genome-positive cell lines, irrespective of whether the cell lines are producers or non-producers of EB virus. Infection of peripheral B lymphocytes of primate origin and continuous EB-virus-DNA-negative B-type cell lines with EB virus results in EBNA formation [5, 27, 96]. Different virus strains behave differently in this respect. The transforming viruses, as B-95-8 virus, induce EBNA in both of these systems, while the non-transforming P3HR-1 virus is only active in the continuous cell lines. It has also been shown that both B-95-8 and P3HR-1 viruses induce EBNA in various types of human and mouse cells onto which EB-virus receptors have been transplanted via reconstituted Sendai virus envelopes [163]. All these results indicate that the non-transforming P3HR-1 virus does not lack the capability of inducing EBNA synthesis, although we do not possess any comparative data proving that EBNA induced by P3HR-1 virus is structurally and functionally identical with that induced by the transforming viruses.

In infected primary cells EBNA appears 12–24h after infection [5, 27, 96, 129, 151]. Provided that sufficient virus concentration is used, more than 90% of these cells may become EBNA-positive at 48h postinfection. Only after appearance of EBNA do the lymphocytes undergo blast formation. At approximately 32–40h after infection, the cells enter S phase, as is evidenced by incorporation of labelled thymidine. This is then followed by cell mitosis and first cell division. Thereafter the appearance of EBNA-positive cells in the culture parallels the growth curve of the transformed cells [65]. It seems that the majority of, if not all, EBNA-positive cells are able to grow [176].

As was observed in the non-producer transformed cells, EBNA is the only viral protein clearly detectable in infected primary cells. However, because of the inhibitory effect of sensitized T lymphocytes on the outcome of the transformation process [16, 126, 153], the synthesis of LYDMA (possibly preceding EBNA formation) seems probable. In the continuous EB-virus-negative lines which have been infected with the non-transforming P3HR-1 virus, the formation of EBNA is followed by EA synthesis in a portion of cells [31, 69, 97].

Although the wealth of the evidence strongly suggests that EBNA formation is a precondition of cell transformation by EB virus, other viral and cell functions are apparently also needed for this process to take place. Different conditions interfering with their expressions may block the development of transformed cell clones. Thus *Thorley-Lawson and Strominger* [154, 155] reported that in the presence of phosphonacetic acid (PPA) (100–200 μg/ml), which is a potent inhibitor of herpesvirus DNA polymerases, infected cells expressed EBNA. For relatively long periods they remained viable and synthesized cell DNA at a constant rate, but they did not start to proliferate until the drug had been removed. When PAA was added at a later stage (3–7 days after infection), it was without effect on cell growth [78, 154]. These observations suggested that viral DNA-polymerase is needed before the transformation phenotype is expressed, possibly because viral DNA replication has to precede its integration into cell DNA. It should be noted that this view is not generally accepted. Results have been obtained suggesting that the drug, at the concentration used by the above authors, exhibited a cytostatic effect and reduced the outgrowth of the transformed cells rather than the transformation itself [126–128]. There are still other observations, the mechanisms of which are even less clear, indicating that EBNA formation is not necessarily followed by cell transformation. One concerns the inhibitory effect of adult fibroblasts on EB-virus-infected lymphocytes [100]. In their presence the infected lymphocytes do not start to proliferate, although EBNA is present in a significant quantity of cells. Addition of the adult fibroblasts to infected lymphocytes 48 h or later after infection does not inhibit cell transformation. It has also been shown that EBNA induction by B-95-8 virus in peripheral leukemic cells is not followed by unlimited proliferation, in most instances not even by blastogenesis and increased DNA synthesis [152].

C. Regulation of EBNA Synthesis

EBNA presence in the transformed cells is conditioned by the growth activity of the cells [148, 169]. In the aging, non-growing cultures the share of EBNA-positive cells decreases more rapidly than the share of living cells [148]. This is different from the antigens related to activated EB-virus cycle, which tend to increase under this condition. It is not yet clear in which stage of the transformed cell-growth cycle EBNA is synthesized. *Ernberg* et al. [29], who followed EBNA synthesis in non-synchronized cells by means of immunofluorometry and measurement of DNA synthesis in individual cells, claimed that the antigen is predominantly formed during the G-1 period. Another group of investigators [138], who used synchronized cells, detected maximal expression of EBNA during early S phase. A positive correlation exists between the number of EB-virus genome equivalents per cell and the extent of EBNA production [28, 137]. This indicated to *Ernberg and Klein* [30] that EBNA formation is an autonomous function of the viral genome expressed independently of other functions.

Heterogeneity of EBNA staining in P3HR-1 virus-converted EB-virus-DNA-negative B-type cell lines has been reported [36, 97]. The biological significance of this finding is unknown. It is likely, however, that the various staining patterns of EBNA in the clones isolated are associated with the heterogeneity of the P3HR-1 virus population [35, 179].

The formation of EBNA in primarily infected B cells can be influenced by various treatments. It can be enhanced by pokeweed mitogen, whereas Concanavalin A is inhibitory and phytohemagglutinin is without effect [77]. It has been reported that interferon does not suppress EBNA formation [98]; however, some more recent results suggest that it can be inhibited by large interferon doses [90]. Inhibitors of protein and RNA synthesis, when added at the time of infection, block EBNA appearance, while DNA synthesis inhibitors somewhat reduce but do not block it [27]. Paradoxically, EBNA formation is inhibited by cytosine arabinoside [67] which may indicate that a low-rate synthesis of DNA, undetectable by other methods, is needed for it. This applies only for primary lymphocytes, however. In the EB-virus-DNA-negative B-cell lines which have been infected with EB virus, EBNA synthesis is not blocked by the substance.

The 'iron rule' of EBNA presence in all EB-virus-genome-positive cells has been disturbed by two findings. The first was the detection of EB-virus-positive EBNA-negative lymphocytes in the blood of IM patients [16, 128]. These cells, characterized by small size and EBNA absence, were capable of producing the transforming virus after having been activated by cultivation in vitro. It has been suggested that these are cells in which the in vivo virus-genome expression has been totally suppressed. However, it remains possible that they represent lymphocytes in the earliest stage of virus infection. The other observation has been reported by *Graessmann* et al. [46]. After microinjecting human fibroblasts and monkey kidney cells with EB-virus DNA they detected EA but not EBNA.

D. Function of EBNA

Most investigators involved in EBNA research, including ourselves, consider EBNA to be an analogue of T antigen of papovaviruses and assume that it is an essential factor in the process of cell transformation and in the maintenance of the transformed phenotype. Some of the main arguments for this view have already been mentioned:

(i) EBNA is universally present in EB-virus-transformed cells;

(ii) in infected B-type cells, the appearance of EBNA precedes the changes associated with transformation [5, 27, 96, 129, 151] and there is a parallelism between the appearance of the EBNA-positive cells and the growth curves of transformed cells [65];

(iii) EBNA is a DNA-binding protein [7, 8, 89]; this is a property which may connect EBNA as a biochemical entity with its proposed regulatory role. As indicated in the preceding section, preliminary evidence suggests a certain degree of specificity of EBNA binding to EB virus DNA [54]; a higher affinity of SV40 T antigen for homologous DNA has also been reported [62, 125, 156, 157].

Recently, another important piece of evidence has been obtained. EBNA has been found to stimulate template activity in vitro [64] and cellular DNA synthesis after microinjection into 3T3 cells [70], somewhat as the T antigen of SV40 does [75, 158].

There are additional observations which highlight the resemblance of EBNA to SV40 T antigen and thus strengthen the possibility that their functions are also similar. These include the recent finding that the two substances adsorb to and elute from cell DNA under the same conditions of pH and ionic strength [146] and also the demonstration that both EBNA and large T antigen combine with proteins (apparently cell proteins) of comparable molecular weight (48K–55K) [70, 76, 86, 94]. Such proteins are also present in cells transformed by some other viruses, by chemical carcinogens, or spontaneously [15, 18]. Most recently they have also been detected in virus-infected cells and in small quantities even in normal non-infected cells [15, 95]. These proteins are species-specific, which suggests that they are host-encoded. Although it is very probable that the complex of the viral and cell products is of high biological significance, its functions are unknown at the moment. Just to mention a few possibilities, the cell protein may stabilize or modify the virus protein essential for transformation, or it may itself exert the function essential for transformation after having been stimulated (and possibly modified) by the corresponding virus product. It is also possible that the viral product and the cell protein cooperate and that their cumulative effects are indispensable for the transformation and its maintenance.

It is to be expected that future studies will reveal even more similarities between EBNA and the T antigens of papova- and adenoviruses. One wonders, for example, whether EBNA will be shown to possess a protein kinase activity (which is apparently possessed by T antigen of SV40 [47, 158]) and whether it will be proven that a split product of EBNA is inserted in the cytoplasmic membrane where it might be an important element of the cell-surface changes associated with the transformation state. The origin and nature of LYDMA may be of special interest in this respect.

Although it is possible that EBNA, like the T antigens, will gradually be shown to be responsible for the multiple functions on DNA replication and its transcription, certain limits have to be imposed on these speculations. The recognition of a multiplicity of T antigens [17, 143] due to RNA splicing has helped to solve the problem of how the many functions can be secured by a single gene product. Still, EBNA is the only virus protein detected in EB-virus-transformed non-producer cells. In the immunoprecipitation studies performed thus far, no clear-cut evidence has been obtained on the presence of the analogues of the 't' or 'mT' antigens. Have they been missed or are they really absent? If they do not exist, this is a significant difference between the two systems under discussion. It should also be kept in mind that there is enough genetic information transcribed in non-producer cells for several medium-sized proteins; some of these may have the functions exerted by the various T antigen species in papovavirus-transformed cells. Because of the large size of the herpesvirus genome, there is much less need—from the point of view of economy—for a single gene product to accomplish all the functions necessary for cell transformation.

In spite of the overwhelming and highly suggestive evidence for a T-antigen-like nature of EBNA,there still remains the possibility that it primarily plays a different role in the transformed cell. The correlation, already mentioned, between the viral DNA content and the amount of EBNA produced [28, 137] may have some biological meaning. For example, it is possible that this is an expression of a repressor-like function of EBNA on virus replication, implying that the more viral DNA is present, the more EBNA is needed. The observation that the increased number of EB-virus genomes results in more easily inducible cells [136, 137] does not necessarily militate against this possibility; the putative balance between EB-virus DNA and EBNA may well become frailer under such conditions.

The studies aiming at a definition of the functions of EBNA are heavily handicapped by two factors. First, monospecific xenogenic antisera against EBNA purified to homogeneity have not yet been prepared. The only reagents available are human sera possessing EBNA antibody. Although they can easily be obtained and checked for antibodies to other known EB-virus-associated antigens, they cannot be examined for antibodies to antigens that have not yet been defined by serological means. The coding capacity of EB-virus DNA exceeds 100 average-sized polypeptides; only a minority has been serologically identified by now. Thus the EBNA-anti-EBNA reaction can be obscured by some other antigen-antibody interaction. This difficulty can be overcome either by preparing high-quality monospecific antisera or by isolating hybridomas producing monoclonal EBNA antibody in vitro.

The other serious difficulty in identifying the transformation functions of EB virus is rooted in the lack of suitable virus mutants which were so rewarding in papovavirus research and in revealing the essential role of their T antigens in transformation [14, 91]. Thermosensitive or deletion mutants with physiological defects influencing the synthesis of EBNA and other viral products would greatly facilitate the understanding of EB-virus biology and dramatically improve the chance of identifying the role which the various virus-specified products play in the establishment and maintenance of the transformation phenotype. Efforts to isolate such EB virus mutants are under way in several laboratories. This work is encumbered with unprecedented difficulties because of the lack of a lytic virus system.

VIII. Summary

Although the viral origin of EBNA has not yet been proven beyond all doubt, present evidence strongly suggests that it is an EB-virus-coded, non-virion antigen. It is present in all EB-virus-transformed cells and can be induced in peripheral B lymphocytes of primate origin and in continuous EB-virus-genome-negative B-type cell lines by infection with EB virus.

The presence of EBNA and antibodies against this antigen can be revealed by the ACIF test on fixed cells and by the CF test with soluble cell extracts. These two tests apparently measure the same activity; however, the ACIF test is more suitable for detecting low levels of antibody. EBNA antibody can be found in nearly all subjects who are VCA-antibody-positive. A notable exception from this rule are IM patients and subjects who have experienced asymptomatic EB-virus infection a short time previously. This is because EBNA antibody appears not earlier than several weeks to several months after infection. EBNA antibody may also be missing in subjects with some inherent immunodeficiencies. In BL and NPC patients, EBNA antibody levels are increased over those determined in sera of normal controls. Repeatedly a relationship between the changes in EBNA-antibody levels and the clinical course of these diseases has been reported. However, it seems clear that changes in EA antibody levels signal the changes in clinical condition more reliably. Elevated titers of EBNA antibody have also been found in several other diseases. While in some of these it is possibly associated with the involvement of EB virus in their etiology, in others it merely reflects extensive virus replication and growth of EBNA-positive clones conditioned by the immunosuppressive effect of the disease itself or the immunosuppressive therapy imposed.

In its native form, EBNA is an oligomeric structure, most probably composed of virus-specific 48K polypeptide complexed with a 53K polypeptide of cellular origin. EBNA binds to DNA, and some very recent results suggest that it possesses a higher affinity for homologous DNA than for at least some heterologous DNAs.

Its biochemical activity and biological functions are not yet understood. However, it presumably plays an important regulatory role in EB-virus-infected cells and may well be the essential factor in their transformation.

Addendum

While this paper was in press, new studies on EBNA were reported. It has been shown by *Purtillo* and his colleagues that patients suffering from X-linked lymphoproliferative syndrome, whose T cells seem to be defective in recognizing EB virus-transformed cells, do not develop anti-EBNA antibody in spite of active and extensive EB virus infection [see the review, *D.T. Purtillo:* 'Immune deficiency to Epstein-Barr virus-induced lymphoproliferative diseases: the X-linked lymphoproliferative syndrome as a model',

Advances in Cancer Research *34:* 279–312, 1981]. This is in line with earlier observations in other inferent immunodeficiences [10]. Using fluoroimmunoelectrophoresis and radioimmunoelectrophoresis, *Strnad* et al. [B.C. Strnad; T.C. Schuster; R.F. Hopkins; R.H. Neubauer; H. Rabin: 'Identification of an Epstein-Barr virus nuclear antigen by fluoroimmunoelectrophoresis and radioimmunoelectrophoresis', J. Virology *38:* 996–1004, 1981] identified 65K protein present in Raji cells (an EB-virus-DNA-positive, non-productive cell line) as an EBNA. Antigens with similar molecular weights were also detected in all three other EB-virus-genome-positive cell lines tested but not in three EB-virus-genome-negative cell lines. These findings are at variance with those reported by *Luka* et al. [86, 88], discussed in this review. It remains to be determined whether the 48K protein detected by *Luka* et al. is a processed product of the 65K protein or whether the 48K and 65K proteins are distinct substances. The demonstration of multiple EBNAs will not be surprising. In the non-productive cell lines the information transcribed from EB-virus DNA into mRNA is sufficient for several medium-sized proteins [66, 115, 116].

Acknowledgments

The authors express their indebtedness to Drs. *D.V. Ablashi, G. de Thé, G. Klein, G. Lenoir, J. Luka, G.R. Pearson* and *J.L. Strominger* for supplying them with valuable materials, including some still-unpublished data on EB virus nuclear antigen.

References

1 Ablashi, D.V.; Easton, J.M.; Armstrong, G.; Bengali, Z.: Extraction of soluble antigens of Epstein-Barr virus, Herpesvirus saimiri, and Herpesvirus ateles with the use of glycine. J. natn. Cancer Inst. *62:* 1173–1175 (1979).

2 Ablashi, D.V.; Gerber, P.; Easton, J.: Oncogenic herpesviruses of non-human primates. Comp. Immunol. Microbiol. infect. Dis. *2:* 229–241 (1979).

3 Andersson-Anvret, M.; Lindahl, T.: Integrated viral DNA sequences on Epstein-Barr virus-converted human lymphoma lines. J. Virol. *25:* 710–713 (1978).

4 Armstrong, D.; Henle, G.; Henle, W.: Complement fixation tests with cell lines derived from Burkitt's lymphoma and acute leukemias. J. Bact. *91:* 1257–1262 (1966).

5 Aya, T.; Osato, T.: Early events in transformation of human cord lymphocytes by Epstein-Barr virus: induction of DNA synthesis, mitosis and the virus-associated nuclear antigen synthesis. Int. J. Cancer *14:* 341–347 (1974).

6 Bahr, G.F.; Mikel, U.; Klein, G.: Localization and quantitation of EBV-associated nuclear antigen (EBNA) in Raji cells. Beitr. Pathol. *155:* 72–78 (1975).

7 Baron, D.; Benz, W.C.; Strominger, J.L.: Assay and partial purification of Epstein-Barr virus nuclear antigen; in Epstein-Barr virus production, concentration and purification. IARC Internal Tech. Rep. No. 75 003, pp. 257–262 (IARC, Lyon 1975).

8 Baron, D.; Strominger, J.L.: Partial purification and properties of the Epstein-Barr virus associated nuclear antigen. J. biol. Chem. *253:* 2875–2881 (1978).

9 Becker, Y.; Weinberg, E.; Cohen, Y.; Klein, G.: Studies on Epstein-Barr virus specified proteins in Burkitt lymphoblasts; in Epstein-Barr virus production, concentration and purification. IARC Internal Tech. Rep. No. 75 003, pp. 285–294 (IARC, Lyon 1975).

10 Berkel, A.I.; Henle, W.; Henle, G.; Klein, G.; Ersoy, F.; Sanal, O.: Epstein-Barr virus related antibody patterns in ataxia-teleangiectasia. Clin. exp. Immunol. *35:* 196–201 (1979).

11 Biggar, R.J.; Henle, G.; Böcker, J.; Lennette, E.T.; Fleisher, G.; Henle, W.: Primary Epstein-Barr virus infection in African infants. II. Clinical and serological observations during seroconversion. Int. J. Cancer *22:* 244–250 (1978)

12 Brown, T.D.K.; Ernberg, I.; Klein, G.: Studies of Epstein-Barr virus (EBV)-associated nuclear antigen. I. Assay in human lymphoblastoid cell lines by direct and indirect determination of ^{125}I-IgG binding. Int. J. Cancer *15:* 606–616 (1975).

13 Brown, T.D.K.; Ernberg, I.; Lamon, E.W.; Klein, G.: Detection of Epstein-Barr virus (EBV)-associated nuclear antigen in human lymphoblastoid cell lines by means of an ^{125}I-IgG binding technique. Int. J. Cancer *13:* 785–794 (1974).

14 Brugge, J.S.; Butel, J.S.: Role of simian virus 40 gene A function in maintenance of transformation. J. Virol. *15:* 619–635 (1975).

15 Carroll, R.B.; Muello, K.; Melero, J.A.: Coordinate expression of the 48K host nuclear phosphoprotein and SV40 T Ag upon primary infection of mouse cells. Virology *102:* 447–452 (1980).

16 Crawford, D.H.; Rickinson, A.B.; Finerty, S.; Epstein, M.A.: Epstein-Barr (EB) virus genome-containing EB nuclear antigen-negative B-lymphocyte populations in blood in acute infectious mononucleosis. J. gen. Virol. *38:* 449–460 (1978).

17 Crawford, V.L.: The T-antigens of simian virus 40 and polyoma virus: their role in transformation. Trends Biochem. Sci. *5:* 39–42 (1980).

18 de Leo, A.B.; Jay, G.; Apella, E.; Dubois, G.C.; Law, L.W.; Old, L.J.: Detection of a transformation-related antigen in chemically induced sarcomas and other transformed cells of the mouse. Proc. natn. Acad. Sci. USA *76:* 2420–2424 (1979).

19 Demissie, A.: Difference in heat stability of antigens associated with Epstein-Barr virus, demonstrated by immunodiffusion. J. natn. Cancer Inst. *51:* 751–760 (1973).

20 de Thé, G.; Day, N.E.; Geser, A.; Lavoué, M.F.; Ho, Y.H.C.; Simons, M.J.; Sohier, R.; Tukei, P.; Vonka, V.; Závadová, H.: Sero-epidemiology of the Epstein-Barr virus: preliminary analysis of an international study – a review; in de Thé, Epstein, zur Hausen, Oncogenesis and herpesviruses II, part 2, pp. 3–16 (IARC, Lyon 1975).

21 de Thé, G.; Geser, A.; Day, N.E.; Tukei, P.M.; Williams, E.H.; Beri, D.P.; Smith, P.G.; Dean, A.G.; Bornkamm, G.W.; Feorino, P.; Henle, G.: Epidemiological evidence for causal relationship between Epstein-Barr virus and Burkitt's lymphoma from Ugandan prospective study. Nature, Lond. *274:* 756–761 (1978).

22 de Thé, G.: Ho, J.H.C.; Ablashi, D.V.; Day, N.E.; Macario, A.J.L.; Martin-Berthelon, M.C.; Pearson, G.; Sohier, R.: Nasopharyngeal carcinoma. IX. Antibodies to EBNA and correlation with response to other EBV antigens in Chinese patients. Int. J. Cancer *16:* 713–721 (1975).

23 de Thé, G.; Sohier, R.; Ho, J.H.C.; Freund, R.: Nasopharyngeal carcinoma. IV. Evolution of complement-fixing antibodies during the course of the disease. Int. J. Cancer *12:* 368–377 (1973).

24 Didier, J.; Dalens, M.; Chabanon, G.; Icart, J.; Enjalbert, L.: The Epstein-Barr virus (EBV) in human pathology. II. Serological profiles of EBV infections. Biomedicine *28:* 54–62 (1978).

25 Dölken, G.; Klein, G.: Radioimmunoassay for Epstein-Barr virus (EBV)-associated nuclear antigen (EBNA). Binding of iodinated antibodies to antigens immobilized in polyacrylamide gel. Eur. J. Cancer *13:* 1227–1236 (1977).

26 Dumont, J.; Liabeuf, A.; Henle, W.; Feingold, N.; Kourilsky, F.M.: Anti-EBV antibody titers in non-Hodgkin lymphomas. Int. J. Cancer *18:* 14–53 (1976).

27 Einhorn, L.; Ernberg, I.: Induction of EBNA precedes the first cellular S-phase after EBV-infection of human lymphocytes. Int. J. Cancer *21:* 157–160 (1978).

28 Ernberg, I.; Andersson-Anvret, M.; Klein, G.; Lundin, L.; Killander, D.: Relationship between amount of Epstein-Barr virus-determined nuclear antigen per cell and number of EBV-DNA copies per cell. Nature, Lond. *266:* 269–270 (1979).

29 Ernberg, I.; Killander, D.; Lundin, L.: Quantity of Epstein-Barr viral nuclear antigen (EBNA) during different phases of the cell cycle (submitted for publication).

30 Ernberg, I.; Klein, G.: EB-virus-induced antigens; in Epstein, Achong, The Epstein-Barr virus, pp. 39–60 (Springer, Berlin 1979).

31 Ernberg, I.; Masucci, G.; Klein, G.: Persistence of Epstein-Barr viral nuclear antigen (EBNA) in cells entering the EB viral cycle. Int. J. Cancer *17:* 197–203 (1976).

32 Falk, L.A.; Deinhardt, F.; Nonoyama, M.; Wolfe, L.G.; Bergolz, C.; Lapin, B.A.; Yakovleva, L.; Agrba, V.; Henle, G.; Henle, W.: Properties of a baboon lymphotropic herpesvirus related to Epstein-Barr virus. Int. J. Cancer *18:* 798–807 (1976).

33 Favart, A.M.; Lamy, M.E.; Burtonboy, G.: Epstein-Barr virus (EBV) intracellular antigens: factors affecting the patterns of immunofluorescence. Archs Virol. *65:* 337–346 (1980).

34 Floyd, R.; Vonka, V.; Benyesh-Melnick, M.: Fluorescence complement fixation by lymphoblastoid cells. J. natn. Cancer Inst. *46:* 383–390 (1971).

35 Fresen, K.O.; Merkt, B.; Bornkamm, G.W.; zur Hausen, H.: Heterogeneity of Epstein-Barr virus originating from P3HR-1 cells. I. Studies on EBNA induction. Int. J. Cancer *19:* 317–323 (1977).

36 Fresen, K.O.; zur Hausen, H.: Establishment of EBNA-expressing cell lines by infection of Epstein-Barr virus (EBV)-genome-negative human lymphoma cells with different EBV strains. Int. J. Cancer *17:* 161–166 (1976).

37 Galas, D.J.; Schmitz, A.: DNAase footprinting: a simple method for the detection of protein-DNA binding specificity. Nucl. Acids Res. *5:* 3158–3170 (1978).

38 Gerber, P.; Birch, S.M.: Complement-fixing antibodies in sera of human and non-human primates to viral antigens derived from Burkitt's lymphoma cells. Proc. natn. Acad. Sci. USA *58:* 478–484 (1967).

39 Gerber, P.; Deal, D.R.: Epstein-Barr virus-induced viral and soluble complement-fixing antigens in Burkitt lymphoma cell cultures. Proc. Soc. exp. Biol. *134:* 748–751 (1970).

40 Gerber, P.; Kalter, S.S.; Schidlovsky, G.; Peterson, W.D., Jr.; Daniel, M.D.: Biological and antigenic characteristics of Epstein-Barr virus-related herpesviruses of chimpanzees and baboons. Int. J. Cancer *20;:* 448–459 (1977).

41 Gerber, P.; Pritchett, R.F.; Kieff, E.: Antigens and DNA of a chimpanzee agent related to Epstein-Barr virus. J. Virol. *19:* 1090–1099 (1976).

42 Gergely, L.; Czeglédy, J.; Szalka, A.; Váczi, L.; Binder, L.: Epstein-Barr virus and cytomegalovirus antibodies in infectious mononucleosis. Acta microbiol. Acad. Sci. hung. *24:* 13–19 (1977).

43 Gilbert, J.A.; Spelsberg, T.C.; Pearson, G.R.: Evidence that class II EBNA is a histone-like protein. Int. Conf. Human Herpesviruses, Atlanta, Ga., 1980, p. G49.

44 Glaser, R.; Ablashi, D.V.; Nonoyama, M.; Henle, W.; Easton, J.: Enhanced oncogenic behavior of human and mouse cells after cellular hybridization with Burkitt tumor cells. Proc. natn. Acad. Sci. USA *74:* 2574–2578 (1977).

45 Glaser, R.; Croce, C.; Nonoyama, M.: Studies on the association of the Epstein-Barr virus genome with chromosomes in human (Burkitt)/mouse hybrid cells; in de Thé, Henle, Rapp, Oncogenesis and herpesviruses III, pp. 565–570 (IARC, Lyon 1978).

46 Graessmann, A.; Wolf, H.; Bornkamm, G.W.: Expression of Epstein-Barr virus genes in different cell types after microinjection of viral DNA. Proc. natn. Acad. Sci. USA *77:* 433–436 (1980).

47 Griffin, J.D.; Spangler, G.; Livingston, D.M.: Protein kinase activity associated with simian virus 40 T antigen. Proc. natn. Acad. Sci. USA *76:* 2610–2614 (1979).

48 Henle, G.; Henle, W.: Immunofluorescence in cells derived from Burkitt's lymphoma. J. Bact. *91:* 1248–1256 (1966).

49 Henle, W.; Henle, G.: Epstein-Barr virus-related serology in Hodgkin's disease. Natn. Cancer Inst. Monogr. *36:* 79–84 (1973).

50 Henle, G.; Henle, W.: The virus as the etiological agent in infectious mononucleosis; in Epstein, Achong, The Epstein-Barr virus, pp. 297–320 (Springer, Berlin 1979).

51 Henle, W.; Henle, G.; Horwitz, C.A.: Antibodies to Epstein-Barr virus-associated nuclear antigen in infectious mononucleosis. J. infect. Dis. *130:* 231–239 (1974).

52 Henle, W.; Ho, J.H.C.; Henle, G.; Chau, J.C.W.; Kwan, H.C.: Nasopharyngeal carcinoma: significance of changes in Epstein-Barr virus-related antibody patterns following therapy. Int. J. Cancer *20:* 663–672 (1977).

53 Hentzen, D.; Lenoir, G.M.; Berthelon, M.-C.; Daillie, J.: Epstein-Barr virus (EBV) antigenic determinants on subunits of the EBV-determined nuclear antigen (EBNA) (submitted for publication).

54 Hesse, J.; Andersen, E.; Levine, P.H.; Ebbesen, P.; Halberg, P.; Reisher, J.I.: Antibodies to Epstein-Barr virus and cellular immunity in Hodgkin's disease and chronic lymphatic leukemia. Int. J. Cancer *11:* 237–243 (1973).

55 Hirsch, I.; Kuchlerová, L.; Břicháček, B.; Suchánková, A.; Vonka, V.: Blocking of acid-fixed nuclear binding of Epstein-Barr virus nuclear antigen (EBNA) by different DNA species. J. gen. Virol. *44:* 849–852 (1979).

56 Hirsch, I.; Suchánková, A.; Závadová, H.; Vonka, V.: Study of Epstein-Barr virus-associated nuclear antigen. Int. J. Cancer *22:* 535–541 (1978).

57 Hirsch, I.; Suchánková, A.; Kuchlerová, L.; Břicháček, B.; Vonka, V.: Interaction of EBNA with anti-EBNA antibody and DNA. Intervirology *13:* 348–351 (1980).

58 Holý, J.; Vaněček, K.; Vonka, V.: Infections with herpesviruses in a closed infants' community. Arch. ges. Virusforsch. *43:* 152–160 (1973).

59 Huang, D.P.; Ho, J.H.C.; Henle, W.; Henle, G.: Demonstration of Epstein-Barr virus-associated nuclear antigen in nasopharyngeal carcinoma cells from fresh biopsies. Int. J. Cancer *14:* 580–588 (1974).

60 Huang, D.P.; Ho, J.H.C.; Henle, W.; Henle, G.; Saw, D.; Lui, M.: Presence of EBNA in nasopharyngeal carcinoma and control patient tissue related to EBV serology. Int. J. Cancer *26:* 266–274 (1978).

61 Jessel, D.; Hudson, J.; Landau, T.; Tenen, D.; Livingston, D.M.: Interaction of partially purified simian virus 40 T antigen with circular viral DNA molecules. Proc. natn. Acad. Sci. USA *72:* 1960–1964 (1975).

62 Jessel, D.; Landau, T.; Hudson, J.; Lalor, T.; Tenen, D.; Livingston, D.M.: Identification of regions of the SV40 genome which contain preferred SV40 T antigen-binding sites. Cell *8:* 535–545 (1976).

63 Johansson, B.; Klein, G.; Henle, W.; Henle, G.: Epstein-Barr virus (EBV)-associated antibody patterns in malignant lymphoma and leukemia. I. Hodgkin's disease. Int. J. Cancer *6:* 450–462 (1970).

64 Kamata, T.; Tanaka, S.; Aikawa, S.; Hunuma, Y.; Watanabe, Y.: A possible function of Epstein-Barr virus-determined nuclear antigen (EBNA): stimulation of chromatin template activity in vitro. Virology *95:* 222–229 (1979).

65 Katsuki, T.; Hinuma, Y.: A quantitative analysis of the susceptibility of human leukocytes to transformation by Epstein-Barr virus. Int. J. Cancer *18:* 7–13 (1976).

66 King, W.; Thomas-Powell, A.L.; Raab-Traub, N.; Hawke, M.; Kieff, E.: Epstein-Barr virus RNA. V. Viral RNA in a restringently infected, growth-transformed cell line. J. Virol. *36:* 506–518 (1980).

67 Klein, G.: Comparison of humoral and cell-mediated responses and virus markers in herpesvirus-associated lymphomas: a review; in de Thé, Henle, Rapp, Oncogenesis and herpesviruses III, pp. 815–833 (IARC, Lyon 1978).

68 Klein, G.; Giovanella, B.C.; Lindahl, T.; Fialkow, P.J.; Singh, S.; Stehlin, J.S.: Direct evidence for the presence of Epstein-Barr virus DNA and nuclear antigen in malignant epithelial cells from patients with poorly differentiated carcinoma of the nasopharynx. Proc. natn. Acad. Sci. USA *71:* 4737–4741 (1974).

69 Klein, G.; Giovanella, B.; Westman, A.; Stehlin, J.S.; Mumford, D.: An EBV-genome-negative cell line established from an American Burkitt lymphoma. Receptor characteristics, EBV infectability and permanent conversion into EBV-positive sublines by in vitro infection. Intervirology *5:* 319–334 (1975).

70 Klein, G.; Luka, J.; Zeuthen, J.: Epstein-Barr virus (EBV)-induced transformation and the role of the nuclear antigen (EBNA). Cold Spring Harb. Symp. quant. Biol. *44:* 253–261 (Cold Spring Harbor Laboratory, Cold Spring Harbor 1979).

71 Klein, G.; Pearson, G.; Nadkarni, J.S.; Klein, E.; Clifford, P.; Henle, G.; Henle, W.: Relationship between Epstein-Barr viral and cell membrane fluorescence of Burkitt tumor cells. I. Dependence of cell membrane fluorescence on presence of EB virus. J. exp. Med. *128:* 1011–1020 (1968).

72 Klein, G.; Svedmyr, E.; Jondal, M.; Persson, P.O.: EBV-determined nuclear antigen (EBNA)-positive cells in the peripheral blood of infectious mononucleosis patients. Int. J. Cancer *17:* 21–26 (1976).

73 Klein, G.; Vonka, V.: Relationship between the Epstein-Barr virus-determined complement-fixing antigen and the nuclear antigen detected by anticomplement immunofluorescence. J. natn. Cancer Inst. *53:* 1645–1646 (1974).

74 Klein, G.; Wiener, F.; Zech, L.; zur Hausen, H.; Reedman, B.: Segregation of the EBV-determined nuclear antigen (EBNA) in somatic cell hybrids derived from the fusion of a mouse fibroblast and a human Burkitt lymphoma line. Int. J. Cancer *14:* 54–64 (1974).

75 Kriegler, M.P.; Griffin, J.D.; Livingston, D.M.: Phenotypic complementation of the SV40 tsA mutant defect in viral DNA synthesis following microinjection of SV40 T antigen. Cell *14:* 983–994 (1978).

76 Lane, D.P.; Crawford, L.V.: T antigen is bound to a host protein in SV40-transformed cells. Nature, Lond. *278:* 261–263 (1979).

77 Leibold, W.; Flanagan, T.D.; Menezes, J.; Klein, G.: Induction of Epstein-Barr virus-associated nuclear antigen during in vitro transformation of human lymphoid cells. J. natn. Cancer Inst. *54:* 65–68 (1975).

78 Lemon, S.M.; Hutt, L.M.; Pagano, J.S.: Epstein-Barr virus transformation of human cord lymphocytes is inhibited by phosphonoacetic acid; in de Thé, Henle, Rapp, Oncogenesis and herpesviruses III, pp. 739–744 (IARC, Lyon 1978).

79 Lenoir, G.; Berthelon, M.C.; Favre, M.C.; de Thé, G.: Characterization of Epstein-Barr virus (EBV) antigens. II. Detection of early antigen(s) using anticomplement immunofluorescence (ACIF) and complement-fixation (CF) tests. Biomedicine *23:* 461–464 (1975).

80 Lenoir, G.; Berthelon, M.C.; Favre, M.C.; de Thé, G.: Characterization of Epstein-Barr virus antigens. I. Biochemical analysis of the complement-fixing soluble antigen and relationship with Epstein-Barr virus-associated nuclear antigen. J. Virol. *17:* 672–674 (1976).

81 Lenoir, G.; de Thé, G.; Ooka, T.; Daillie, J.: Studies on Epstein-Barr virus early proteins. Coll. INSERM *69:* 165–172 (1977).

82 Lenoir, G.; de Thé, G.; Virelizier, J.L.; Griscelli, C.: Epstein-Barr virus nuclear antigen (EBNA)-positive cells in a lymph node of a child with severe primary EBV infection. In de Thé, Henle, Rapp, Oncogenesis and herpesviruses III, pp. 733–738 (IARC, Lyon 1978).

83 Levine, P.H.; Ablashi, D.V.; Berard, C.W.; Carbone, P.P.; Waggoner, D.E.; Malan, L.: Elevated antibody titers to Epstein-Barr virus in Hodgkin's disease. New Engl. J. Med. *24:* 181–186 (1974).

84 Levine, P.H.; Wallen, W.C.; Ablashi, D.V.; Granlund, D.J.; Connelly, R.: Comparative studies on immunity to EBV-associated antigens in NPC patients in North America, Tunisia, France and Hong Kong. Int. J. Cancer *20:* 332–338 (1977).

85 Lindahl, T.; Klein, G.; Reedman, B.M.; Johansson, B.; Singh, S.: Relationship between Epstein-Barr virus-determined nuclear antigen (EBNA) in Burkitt lymphoma biopsies and other lymphoproliferative malignancies. Int. J. Cancer *13:* 764–772 (1974).

86 Luka, J.; Jörnvall, H.; Klein, G.: Purification and biochemical characterization of the Epstein-Barr virus (EBV)-determined antigen (EBNA) and an associated protein with a 53K subunit. J. Virol. *35:* 592–602 (1980).

87 Luka, J.; Klein, G.; Henle, G.; Henle, W.: Detection of the EBV-determined nuclear antigen (EBNA) in Burkitt's lymphoma and nasopharyngeal carcinoma biopsies by the acid-fixed nuclear binding (AFNB) technique. Cancer Lett. *4:* 199–205 (1978).

88 Luka, J.; Lindahl, T.; Klein, G.: Purification of the Epstein-Barr virus-determined nuclear antigen from Epstein-Barr virus-transformed human lymphoid cell lines. J. Virol. *27:* 604–611 (1978).

89 Luka, J.; Siegert, W.; Klein, G.: Solubilization of the Epstein-Barr virus-determined nuclear antigen and its characterization as a DNA-binding protein. J. Virol. *22:* 1–8 (1977).

90 Lvovsky, E.; Levine, P.H.; Fucillo, D.; Ablashi, D.; Bengali, Z.H.; Armstrong, G.R.; Levy, H.B.: Epstein-Barr virus and herpesvirus saimiri: sensitivity to interferons and interferon inducers. J. natn. Cancer Inst. (in press).

91 Martin, R.G.; Chou, J.Y.: Simian virus 40 functions required for the establishment and maintenance of malignant transformation. J. Virol. *15:* 619–635 (1971).

92 Matsuo, T.; Hibi, N.; Nishi, S.; Hirai, H.; Osato, T.: Studies on Epstein-Barr virus-related antigens. III. Purification of the virus-determined nuclear antigen (EBNA) from nonproducer Raji cells. Int. J. Cancer *22:* 747–752 (1978).

93 Matsuo, T.; Nishi, S.; Hirai, H.; Osato, T.: Studies of Epstein-Barr virus-related antigens. II. Biochemical properties of soluble antigen in Raji Burkitt lymphoma cells. Int. J. Cancer *19:* 364–370 (1977).

94 McCormick, F.; Harlow, E.: Association of a murine 53,000-dalton phosphoprotein with simian virus 40 large-T antigen in transformed cells. J. Virol. *34:* 213–224 (1980).

95 Mellero, J.A.; Tur, S.; Carroll, R.B.: Host nuclear proteins expressed in simian virus 40-transformed and -infected cells. Proc. natn. Acad. Sci. USA *77:* 97–101 (1980).

96 Menezes, J.; Jondal, M.; Leibold, W.; Dorval, G.: Epstein-Barr virus interaction with human lymphocyte subpopulations: virus adsorption, kinetics of expression of EBV-associated nuclear antigen (EBNA) and lymphocyte transformation. Infect. Immunity *13:* 303–310 (1976).

97 Menezes, J.; Patel, P.; Dussault, H.; Bourkas, A.E.: Comparative studies on the induction of virus-associated nuclear antigen and early antigen by lymphocyte-transforming (B95-8) and non-transforming (P3HR-1) strains of Epstein-Barr virus. Intervirology *9:* 86–94 (1978).

98 Menezes, J.; Patel, P.; Dussault, M.; Joncas, J.; Leibold, W.: Effect of interferon on lymphocyte transformation and nuclear antigen production by Epstein-Barr virus. Nature, Lond. *260:* 430–432 (1976).

99 Miller, G.; Shope, T.; Coope, D.; Waters, L.; Pagano, J.; Bornkamm, G.W.; Henle, W.: Lymphoma in cotton-top marmosets after inoculation with Epstein-Barr virus: tumor incidence, histologic spectrum, antibody responses, demonstration of viral DNA, and characterization of viruses. J. exp. Med. *145:* 948–967 (1977).

100 Moss, D.J.; Pope, J.H.; Scott, W.: Inhibition of EB virus transformation of non-adherent human lymphocytes by co-cultivation with adult fibroblasts. Med. Microbiol. Immunol. *162:* 159–167 (1976).

101 Ng, H.H.; Shortridge, K.F.; Ng, W.S.; Kwan, H.C.: Studies on the isolation of cell-associated Epstein-Barr virus antigens by affinity chromatography; in de Thé, Epstein, zur Hausen, Oncogenesis and herpesviruses II, pp. 277–284 (IARC, Lyon 1975).

102 Nkrumah, F.; Henle, W.; Henle, G.; Herberman, R.B.; Perkins, V.; Depue, R.: Burkitt's lymphoma: its clinical course in relation to immunological reactivities to Epstein-Barr virus and tumor-related antigens. J. natn. Cancer Inst. *57:* 1051–1056 (1976).

103 Nonoyama, M.; Pagano, J.: Complementary RNA specific to the DNA of Epstein-Barr virus: detection of EB viral genomes in nonproductive cells. Nature new Biol. *233:* 103–106 (1971).

104 Oettgen, H.F.; Aoki, T.; Geering, G.; Boyse, E.A.; Old, L.J.: Definition of an antigenic system associated with Burkitt lymphoma. Cancer Res. *28:* 1288—1299 (1968).

105 Ohno, S.; Luka, J.; Falk, L.; Klein, G.: Detection of a nuclear, EBNA-type antigen in apparently EBNA-negative Herpesvirus papio(HVP)-transformed lymphoid lines by the acid-fixed nuclear binding technique. Int. J. Cancer *20:* 941–946 (1977).

106 Ohno, S.; Luka, J.; Falk, L.A.; Klein, G.: Serological reactivities of human and baboon sera against EBNA and Herpesvirus papio-determined nuclear antigen. Eur. J. Cancer *14:* 955–960 (1978).

107 Ohno, S.; Luka, J.; Klein, G.: Evidence for antigenic distinctness of the Epstein-Barr

virus-determined nuclear antigen and the Herpesvirus papio-determined nuclear antigen. Cancer Lett. *6:* 325–329 (1979).

108 Ohno, S.; Luka, J.; Klein, G.; Daniel, M.D.: Detection of a nuclear antigen in *Herpesvirus ateles*-carrying marmoset lines by the acid-fixed nuclear binding (AFNB) technique. Proc. natn. Acad. Sci. USA *76:* 2042–2046 (1979).

109 Ohno, S.; Luka, J.; Lindahl, T.; Klein, G.: Identification of a purified complement-fixing antigen as the Epstein-Barr virus-determined nuclear antigen (EBNA) by its binding to metaphase chromosomes. Proc. natn. Acad. Sci. USA *74:* 1605–1609 (1977).

110 Ohno, S.; Wiener, F.; Klein, G.: Histochemical studies on the EBV-determined nuclear antigen (EBNA). Biomedicine *26:* 268–275 (1977).

111 Orellana, T.; Kieff, E.: Epstein-Barr virus-specific RNA. II. Analysis of polyadenylated viral RNA in restringent, abortive and productive infections. J. Virol. *22:* 321–330 (1977).

112 Pikler, G.M.; Pearson, G.R.; Spelsberg, T.C.: Isolation of the Epstein-Barr virus nuclear antigen from chromatin preparations; in de Thé, Henle, Rapp, Oncogenesis and herpesviruses III, pp. 243–247 (IARC, Lyon 1978).

113 Pikler, G.M.; Spelsberg, T.C.; Pearson, G.R.: The Epstein-Barr virus nuclear antigen. Cell Nucleus *7:* 563–579 (1979).

114 Pope, J.H.; Horne, M.K.; Wetters, E.J.: Significance of a complement-fixing antigen associated with herpes-like virus and detected in the Raji cell line. Nature, Lond. *222:* 166–167 (1969).

115 Powell, A.L.T.; King, W.; Kieff, E.: Epstein-Barr virus-specific RNA. III. Mapping of DNA encoding viral RNA in restringent infection. J. Virol. *29:* 261–275 (1979).

116 Pritchett, R.; Pedersen, M.; Kieff, E.: Complexity of EBV homologous DNA in continuous lymphoblastoid cell lines. Virology *74:* 227–231 (1976).

117 Prives, C.; Gilboa, E.; Revel, M.; Winocour, E.: Cell-free translation of simian virus 40 early messenger RNA coding for viral T-antigen. Proc. natn. Acad. Sci. USA *74:* 457–461 (1977).

118 Ptashne, M.; Jeffrey, A.; Johnson, A.D.; Maurer, R.; Meyer, B.J.; Pabo, C.O.; Roberts, T.M.; Sauer, R.T.: How the λ repressor and cro work. Cell *19:* 1–11 (1980).

119 Rabin, H.; Strnad, B.C.; Neubauer, R.H.; Brown, A.M.; Hopkins, R.F., III.; Mazur, R.A.: Comparisons of nuclear antigens of Epstein-Barr virus (EBV) and EBV-like simian viruses. J. gen. Virol. *48:* 265–272 (1980).

120 Rabin, H.; Neubauer, R.H.; Hopkins, R.F., III.; Dzhikidze, E.K.; Shevtsova, Z.V.; Lapin, B.A.: Transforming activity and antigenicity of an Epstein-Barr-like virus from lymphoblastoid cell lines of baboons with lymphoid disease. Intervirology *8:* 240–249 (1977).

121 Reed, S.I.; Ferguson, J.; Davis, R.W.; Stark, G.R.: T antigen binds to simian virus 40 DNA at the origin of replication. Proc. natn. Acad. Sci. USA *72:* 1605–1609 (1975).

122 Reedman, B.M.; Klein, G.: Cellular localization of an Epstein-Barr virus (EBV)-associated complement-fixing antigen in producer and non-producer lymphoblastoid cell lines. Int. J. Cancer *11:* 499–520 (1973).

123 Reedman, B.M.; Klein, G.; Pope, J.H.; Walters, M.K.; Hilgers, J.; Singh, S.; Johansson, B.: Epstein-Barr virus-associated complement fixing and nuclear antigens in Burkitt lymphoma biopsies. Int. J. Cancer *13:* 755–763 (1974).

124 Reedman, B.M.; Pope, J.H.; Moss, D.J.: Identity of the soluble EBV-asasociated antigens of human lymphoid cell lines. Int. J. Cancer *9:* 172–181 (1972).

125 Reiser, J.; Renart, J.; Crawford, L.V.; Stark, G.R.: Specific association of simian virus 40 tumor antigen with simian virus 40 chromatin. J. Virol. *33:* 78–87 (1980).

126 Rickinson, A.B.; Crawford, D.; Epstein, M.A.: Inhibition of the in vitro outgrowth of Epstein-Barr virus-transformed lymphocytes by thymus-dependent lymphocytes from infectious mononucleosis patients. Clin. exp. Immunol. *28:* 72–79 (1977).

127 Rickinson, A.B.; Epstein, M.A.: Sensitivity of the transforming and replicative functions of Epstein-Barr virus to inhibition by phosphonoacetate. J. gen. Virol. *40:* 409–420 (1978).

128 Rickinson, A.B.; Finerty, S.; Epstein, M.A.: Mechanism of the establishment of Epstein-Barr virus genome-containing lymphoid cell lines from infectious mononucleosis patients: studies with phosphonoacetate. Int. J. Cancer *20:* 861–868 (1977).

129 Robinson, J.; Miller, G.: Assay for Epstein-Barr virus based on stimulation of DNA synthesis in mixed leukocytes from human umbilical cord. J. Virol. *15:* 1065–1071 (1975).

130 Rocchi, G.; de Felici, A.; Ragona, G.; Heinz, A.: Quantitative evaluation of Epstein-Barr-virus-infected mononuclear peripheral blood leukocytes in infectious mononucleosis. New Engl. J. Med. *296:* 132–134 (1977).

131 Rocchi, G.; Hewetson, J.; Henle, W.: Specific neutralizing antibodies in Epstein-Barr virus-associated diseases. Int. J. Cancer *11:* 637–647 (1973).

132 Rocchi, G.; Tosato, G.; Papa, G.; Ragona, G.: Antibodies to Epstein-Barr virus-associated nuclear antigen and to other viral and non-viral antigens in Hodgkin's disease. Int. J. Cancer *16:* 323–328 (1975).

133 Rosén, A.; Jansson, I.; Luka, J.; Klein, G.: A new micro-ELISA method for detection of antibodies against EBV-associated nuclear antigens (submitted for publication).

134 Rymo, L.: Identification of transcribed regions of Epstein-Barr virus DNA in Burkitt lymphoma-derived cells. J. Virol. *32:* 8–18 (1979).

135 Shamoto, M.; Suzuki, I.: An immunoelectron microscopic analysis of Epstein-Barr virus-associated complement-fixing antigen. Cancer, N.Y. *38:* 2057–2064 (1976).

136 Shapiro, I.; Andersson-Anvret, M.; Klein, G.: Polyploidization of Epstein-Barr virus(EBV)-carrying lymphoma lines decreases the inducibility of EBV-determined early antigen following P3HR-1 virus superinfection or iododeoxyuridine treatment. Intervirology *10:* 94–101 (1978).

137 Shapiro, I.M.; Luka, J.; Andersson-Anvret, M.; Klein, G.: Relationship between Epstein-Barr virus genome number, amount of nuclear antigen, and early antigen inducibility in diploid and tetraploid lymphoma cells of related origin. Intervirology *12:* 19–25 (1979).

138 Slovin, S.F.; Vaughan, J.H.; Carson, D.A.: Changes in the expression of two Epstein-Barr virus-associated antigens, EBNA and RANA, during the cell cycle of transformed human B lymphoblasts. Int. J. Cancer *26:* 9–12 (1980).

139 Soběslavský, O.; Kouba, K.; Lasovská, J.; Pokorný, J.; Kucharská, Z.; Vojtěchovský, K.; Seeman, J.; Syrůček, L.: Incidence of antibodies to EB virus in infectious mononucleosis patients. J. Hyg. Epidemiol. Immunol. *16:* 370–376 (1972).

140 Sohier, R.; de Thé, G.: Fixation du complément avec un antigéne soluble: différences d'activité importantes entre les sérums de lymphome de Burkitt, de cancer du rhinopharynx et de mononucléose infectieuse. C. r. hebd. Séanc. Acad. Sci., Paris *273:* 121–124 (1971).

141 Sohier, R.; de Thé, G.: Evolution of complement-fixing antibody titres with the development of Burkitt's lymphoma. Int. J. Cancer *9:* 524–528 (1972).

142 Sohier, R.; de Thé, G.: EBNA antibodies are more useful than complement fixing antibodies in monitoring infectious mononucleosis patients. Biomedicine *29:* 170–173 (1978).

143 Spangler, G.J.; Griffin, J.D.; Rubin, H.; Livingston, D.M.: Identification and initial characterization of a new low-molecular-weight virus-encoded T antigen in a line of simian virus 40-transformed cells. J. Virol. *36:* 488–498 (1980)

144 Spira, J.; Povey, S.; Wiener, F.; Klein, G.; Andersson-Anvret, M.: Chromosome banding, isoenzyme studies and determination of Epstein-Barr virus DNA content on human Burkitt lymphoma/mouse hybrids. Int. J. Cancer *20:* 849–858 (1977).

145 Stevens, D.A.; Pry, T.W.; Blackham, E.A.: Prevalence of precipitating antibody to antigens derived from Burkitt lymphoma cultures infected with herpes-type virus (EB-virus). Blood *35:* 263–275 (1970).

146 Suchánková, A.; Hirsch, I.; Vonka, V.: Binding of SV40 tumor antigen and Epstein-Barr-virus nuclear antigen to isolated acid-fixed nuclei. Acta virol., Prague *24:* 114–118 (1980).

147 Sutton, R.N.P.; Rēynolds, K.; Almond, J.P.; Marston, S.D.; Edmond, R.T.D.: Immunoglobulins and EB virus antibodies in infectious mononucleosis. Clin. exp. Immunol. *13:* 359–366 (1973).

148 Suzuki, M.; Hinuma, Y.: Relationship between Epstein-Barr virus-associated nuclear antigen and cell viability in human cell lines. Gann *66:* 103–105 (1975).

149 Svedmyr, E.; Jondal, M.: Cytotoxic cells specific for B cell lines transformed by Epstein-Barr virus are present in patients with infectious mononucleosis. Proc. natn. Acad. Sci. USA *72:* 1622–1626 (1975).

150 Szigeti, R.; Luka, J.; Klein, G.: Leucocyte migration inhibition studies with Epstein-Barr virus (EBV)-determined nuclear antigen (EBNA) in relation to the EBV-carrier status of the donor. Cell. Immunol.. in press (1981).

151 Takada, K.; Osato, T.: Analysis of the transformation of human lymphocytes by Epstein-Barr virus. I. Sequential occurrence from the virus-determined nuclear antigen synthesis, to blastogenesis, to DNA synthesis. Intervirology *11:* 30–39 (1978).

152 Takada, K.; Yamamoto, K.; Osato, T.: Analysis of the transformation of human lymphocytes by Epstein-Barr virus. II. Abortive response of leukemic cells to the transforming virus. Intervirology *13:* 223–231 (1980)

153 Thorley-Lawson, D.A.; Chess, L.; Strominger, J.L.: Suppression of in vitro Epstein-Barr virus infection: a new role for adult T-lymphocytes; in de Thé, Henle, Rapp, Oncogenesis and herpesviruses III, pp. 623–626 (IARC, Lyon 1978).

154 Thorley-Lawson, D.A.; Strominger, J.L.: Transformation of human lymphocytes by Epstein-Barr virus is inhibited by phosphonoacetic acid. Nature, Lond. *263:* 332–334 (1976).

155 Thorley-Lawson, D.A.; Strominger, J.L.: Reversible inhibition by phosphonoacetic acid of human B lymphocyte transformation by Epstein-Barr virus. Virology *86:* 422–431 (1978).

156 Tjian, R.: The binding site on SV40 DNA for a T antigen-related protein. Cell *13:* 165–179 (1978).

157 Tjian, R.; Fey, G.; Graessmann, A.: Biological activity of purified simian virus 40 T antigen proteins. Proc. natn. Acad. Sci. USA *75:* 1279–1283 (1978).

158 Tjian, R.; Robbins, A.: Enzymatic activities associated with a purified simian virus 40 T-antigen-related protein. Proc. natn. Acad. Sci. USA *76:* 610–614 (1979).

159 Tsang, K.Y.; Hann, W.D.: Activation of Epstein-Barr virus hybrid cells. J. natn. Cancer Inst. *58:* 1295–1299 (1977).

160 Twardzik. R.D.; Armstrong, G.R.; Ablashi, D.V.: Multiple soluble complement-fixing antigens in an Epstein-Barr virus (EBV)-producing cell line; in Epstein-Barr virus production, concentration and purification, IARC Internal Tech. Rep. No. 75003, pp. 245–256 (IARC, Lyon 1975).

161 Veltri, R.W.; Heyl, L.W.; Sprinkle, P.M.: Epstein-Barr virus genome-carrying lymphocyte subpopulations of human palatine tonsils. Proc. Soc. exp. Biol. Med. *156:* 282–286 (1977).

162 Veltri, R.J.; McClung, J.; Sprinkle, P.: Epstein-Barr nuclear antigen (EBNA)-carrying lymphocytes in human palatine tonsils. J. gen. Virol. *32:* 455–460 (1976).

163 Volsky, D.J.; Shapiro, I.M.; Klein, G.: Transfer of Epstein-Barr virus receptors to receptor-negative cells permits virus penetration and antigen axpression. Proc. natn. Acad. Sci. USA *77:* 5453–5457 (1980).

164 Vonka, V.; Benyesh-Melnick, M.: EB virus genome expressions. Colloquium on Oncogenic Viruses. Xth Int. Congr. for Microbiology, Mexico, 1970.

165 Vonka, V.; Benyesh-Melnick, M.; Lewis, R.T.; Wimberly, I.: Some properties of the soluble (S) antigen of cultured lymphoblastoid cell lines. Arch. ges. Virusforsch. *31:* 113–124 (1970).

166 Vonka, V.; Benyesh-Melnick, M.; McCombs, R.M.: Antibodies in human sera to soluble and viral antigens found in Burkitt lymphoma and other lymphoblastoid cell lines. J. natn. Cancer Inst. *44:* 865–872 (1970).

167 Vonka, V.; Porter, D.D.; McCombs, R.M.; Benyesh-Melnick, M.: Studies with viral and soluble antigens from cultured Burkitt lymphoma cells. Bact. Proc., p. 154 (1969).

168 Vonka, V.; Vlčková, I.; Závadová, H.; Kouba, K.; Lazovská, J.; Duben, J.: Antibodies to EB virus capsid antigen and to soluble antigen of lymphoblastoid cells in infectious mononucleosis patients. Int. J. Cancer *9:* 529–535 (1972).

169 Vonka, V.; Šíbl, O.; Suchánková, A.; Simonová, I.; Závadová, H.: Epstein-Barr virus antibodies in tonsillar carcinoma patients. Int. J. Cancer *19:* 456–459 (1977).

170 Wainwright, W.; Veltri, R.: Separation of the complement-fixing and early antigens from Epstein-Barr soluble antigen. J. natn. Cancer Inst. *58:* 1111–1113 (1977).

171 Wallen, W.C.; Mattson, J.M.; Levine, P.H.: Detection of soluble antigen of Epstein-Barr virus by the enzyme-linked immunosorbent assay. J. infect. Dis. *136:* S 324–S 328 (1977).

172 Walters, M.K.; Pope, J.H.: Studies of the EB virus-related antigens of human leucocyte lines. Int. J. Cancer *8:* 32–40 (1971).

173 Weliky, N.; Leaman, D.H., Jr.; Kalman, B.H.: Purification of soluble complement fixing antigens from two Burkitt lymphoma cell lines. Immunology *29:* 779–790 (1975).

174 Yata, J.; Desranges, C.; Nakagawa, T.; Favre, M.C.; de Thé, G.: Lymphoblastoid transformation and kinetics of appearance of viral nuclear antigen (EBNA) in cord blood lymphocytes infected by Epstein-Barr virus (EBV). Int. J. Cancer *15:* 377–384 (1975).

175 Young, R.C.; Corder, M.P.; Haynes, H.A.; Vita, V.T.: Delayed hypersensitivity in Hodgkin's disease. A study of 103 untreated patients. Am. J. Med. *52:* 63–72 (1972).

176 Zerbini, M.; Ernberg, I.: The Epstein-Barr virus (EBV) infection and growth-stimulating effect in human B-lymphocytes. J. gen. Virol., in press (1981).

177 zur Hausen, H.; Fresen, K.O.: Heterogeneity of Epstein-Barr virus. II. Induction of early antigens (EA) by complementation. Virology *81:* 138–143 (1977).

178 zur Hausen, H.; Fresen, K.O.: Heterogeneity of Epstein-Barr virus. Biochim. biophys. Acta *560:* 343–353 (1979).

179 zur Hausen, H.; Schulte-Holthausen, H.: Presence of EB virus nucleic acid in a 'virus-free' line of Burkitt tumor cells. Nature, Lond. *227:* 245–248 (1970).

Dr. Vladimir Vonka, Department of Experimental Virology, Institute of Sera and Vaccines, W. Pieck Street 108, Praha 10 (Czechoslovakia)

Prog. med. Virol., vol. 28, pp. 180–191 (Karger, Basel 1982)

Benefits, Risks and Costs of Viral Vaccines

Jeffrey P. Koplan, Norman W. Axnick[1]

Office of Program Planning and Evaluation, Centers for Disease Control, Atlanta, Ga., USA

Contents

Introduction

A quantitative, albeit arbitrary and pre-metric, formula with a benefit-cost ratio of 16:1 was stated in 'Poor Richard's Almanac' by *Benjamin Franklin* who equated 'an ounce of prevention' with 'a pound of cure'. Beyond folk aphorisms, the concept of the cost-effectiveness of preventive measures in viral illness is not just of recent interest, as evidenced in the January 1879 *Scientific American:*

> 'Loss of life by yellow fever in the South last year is estimated at about 15,000 persons, and of money and trade at from $ 175,000,000 to $ 200,000,000; as great as the loss from the Chicago fire. But some good is likely to come out of the calamity. It is thought that henceforth quarantine regulations will be more thoroughly established than they have ever been. Apart from death and human suffering, negligence is the worst kind of economy. Expenditure of a twentieth part of what the fever has cost might have prevented it altogether.'

[1] We thank Dr. *Alan Hinman* for his critical review of the manuscript and Mrs. *Renee Shirley* for her excellent work in its preparation.

Disease prevention is practiced with the implicit understanding that it is worthwhile in and of itself. However, the costs of prevention are also compared with the costs of patient management and treatment. Our curative armamentarium is limited for most viral infections, and the ability to prevent these infections with vaccinations is highly desirable, particularly when they are common and/or incur great losses by morbidity or mortality. Assessment of a given viral vaccine involves weighing its risks and benefits, with costs often considered as well.

Vaccine risks are the untoward reactions of vaccination. Vaccine benefits are the prevented morbidity and mortality of the disease for which vaccine is being given. The direct costs of vaccination include those of the vaccine and its administration and those incurred by adverse reactions. The direct costs of disease include the costs of medical care for cases of uncomplicated illness and those for cases with complications or sequelae, as well as educational and long-term disability costs. Some studies also consider indirect costs which include the cost of time lost from work due to a physician visit or illness and lost lifetime earnings resulting from disability and premature death. In addition, net costs (and sometimes benefits) are frequently 'discounted' [1–3]. As *Shephard and Thompson* [2] explain:

> 'Discounting is a procedure economists use to relate costs and savings occurring at different times to a common basis. The principle is that future costs are less expensive than present costs because (a) most people would accept less money to receive it sooner and (b) a smaller amount of money can be invested by society and allowed to grow at a compound rate of interest (analogous to the growth of a savings account) to yield the amount of money required for future costs. This rate of interest is called the discount rate... usually in the range of 5 to 15 percent... The discount rate used should reflect opportunity costs and the rate of time preference for money.'

Three types of formal analyses are usually done to evaluate viral vaccines: *benefit-risk analysis, benefit-cost analysis* and *cost-effectiveness analysis* [1–3]. Benefit-risk studies compare vaccination with not vaccinating in terms of absolute numbers or rates of morbidity and mortality. Many studies titled 'benefits and risks' provide descriptive rather than quantitative information which is necessary for formal benefit-risk analysis. In benefit-cost studies, all benefits to society and risks are usually expressed in economic terms. Total costs of vaccination are compared to total costs of disease with and without vaccination and expressed as the difference in benefits and costs (net benefits). In cost-effectiveness studies, vaccination costs are determined and then expressed in terms of what is gained, e.g., per case of disease or death averted per year of productive life gained, etc.

In our view, the value of these types of analysis lies not in providing the definitive basis for a decision on vaccine use or vaccine evaluation but in providing a structured framework which permits the decision-maker to consider all relevant components of the decision in perspective to their relative contributions and the relevant effects of the decision. It forces key assumptions to be made explicit and identifies areas in which data are inadequate. The results of a benefit-cost or cost-effectiveness analysis on a viral vaccine can assist in justifying a traditional approach to scientific advancement (poliomyelitis [4]), assist in disseminating a vaccination program more widely (measles [5]), or assist in a major change in health policy (smallpox [6]). We will review benefit-risk, benefit-cost and cost-effectiveness analyses of several viral vaccines.

Smallpox

Active immunization against smallpox is one of medicine's oldest and most effective preventive efforts. An informal weighing of risks and benefits in the immunizing process has been done in many societies over hundreds of years. Prior to *Jenner's* introduction of vaccination with cowpox virus, variolation was the only means of immunization and was widely practiced in China, India, Turkey and Africa prior to its introduction to England by *Lady Mary Wortley Montagu* in the early 18th century [7]. The benefit-risk assessment implicit in variolation was that application of infectious smallpox material (scab or fluid) by inhalation, skin abrasion or puncture produced a milder form of illness vwith localized lesions and less likelihood of severe disfigurement or death than natural infection.

Immunization with vaccinia virus avoided the risks of variolation – in particular, developing florid smallpox from the procedure – and provided the full benefits of protection against natural infection. The benefits of vaccination in protecting individuals against a widespread and serious disease, when considered along with the rarity of serious adverse reactions to vaccinations, made smallpox vaccination in the United States a medical procedure of unquestioned value until the mid-1960s. The last case of variola minor in the United States was in 1949. By 1971, with the efforts of the WHO Smallpox Eradication Program, only 9 countries remained endemic, all in Asia and Africa.

Given these circumstances, *Lane, Neff,* and co-workers sought to determine the incidence of adverse reactions to smallpox vaccination and determine whether the risks of immunization outweighed the greatly diminished risks of being exposed to smallpox. Based on national and state-wide surveys done in 1963 and 1968 [8–12], the risk of death from all smallpox vaccination complications was 1.0 per million for primary vaccinees of all ages and 0.1 per million for revaccinees of all ages. Among primary vaccinees the risk of death from all complications was 5.0 per million primary vaccinations for children under the age of 12 months and 0.5 per million primary vaccinations for persons 1 through 19 years of age. In addition, among primary vaccinees, the combined rate of post-vaccinial encephalitis and vaccinia necrosum was 6.5 per million for infants and 3.0 per million for persons aged 1 through 19 years.

The risks of smallpox exposure and infection were considered to be the probability of smallpox importation into the United States and the extent of spread after importation. Based on a survey of travelers, it was estimated that the probability of a smallpox importation into the USA in 1970 was 0.0828 or one importation every 12 years. European experience in smallpox spread after importation suggested that it would require 15 smallpox importations per year to produce the same mortality which was then associated with smallpox vaccination in the United States [6, 13].

The result of these studies was a major alteration in national health policy – the recommendation that routine smallpox vaccination be discontinued [6]. *Sencer and Axnick* [14] estimated that the costs of routine smallpox vaccination in 1968 were 159 million dollars in the United States, representing the costs of vaccination, time lost from work, physician fees, and the treatment of vaccination complications. With the discontinuation of routine smallpox vaccination, much of this expenditure could be realized as savings.

Polio

The staggering burden of poliomyelitis was epitomized by the equipment required to support the survivors of paralytic disease – the wards filled with respirators, and the dispensaries stocked with wheel chairs, braces, and crutches. Even without formal benefit-cost or cost-effectiveness analysis, it is apparent that vaccine against poliomyelitis has enabled society to save many dollars, and avoid considerable death, disease, and suffering.

Weisbrod [4] used benefit-cost analysis to consider the economic impact of poliomyelitis vaccine, including research expenditures leading to the development of the vaccine. Using direct and indirect costs of vaccination and research compared to the costs of disease, vaccine benefits are expressed as rates of return which vary from 0.4 to 14.2% depending on the assumptions made. For benefits extending from 1930 through 2200, assuming vaccine costs of 22 cents per dose, the rates of return range from 7.0 to 14.2%.

A methodologic problem of this analysis, acknowledged by the author, lies in determining which basic research contributed to the development of poliomyelitis vaccine – some funding awards listed as fellowships or grants from the National Foundation for Infantile Paralysis were given for research in basic virology and molecular biology and some investigations, supported by funding sources which might not list the research as directed towards poliomyelitis, might have contributed to the vaccine development. Similarly, research leading to the development of the vaccine (e.g., the Nobel Prize winning work in virology of, *Enders, Robbins* and *Weller*) has had major scientific benefits beyond the development of poliomyelitis vaccine.

There is an additional problem of definitions in the analysis. Cost figures for each case of poliomyelitis represent a weighted average of paralytic and non-paralytic cases ($ 1,150 mean loss per case). However, the incidence figures, used to determine the number of cases of disease that are prevented by vaccination, are reported figures from a national surveillance system, in which paralytic cases are far more likely to be reported than non-paralytic cases. Thus the cost figures for disease may well be low estimates and the benefits of vaccination even greater than this study suggests.

In a similar study, including basic research costs in a benefit-cost analysis of poliomyelitis vaccine, *Fudenberg* [15] expressed the results in estimated total dollars saved rather than as a percentage rate of return. Comparing the 6 years pre-vaccination to the 6 years following polio vaccine introduction, the savings in prevented disease (vaccine benefits) are listed as $ 326.8 million in direct costs and $ 6,389.7 million in indirect costs, totalling $ 6,716.5 million. The costs of a vaccination program total $ 611.7 million including vaccine, physician fees and vaccine administration. If the costs of the poliomyelitis vaccine research and field trials are added ($ 41.3 million), the total expenditure is $ 653 million. When considered over the 6-year period, 1955–1961, it is estimated that the net benefits were $ 1 billion per year. It is unclear whether vaccination costs include the indirect costs of time lost from work for working adults to take themselves or their children for vaccination.

In addition, the true basic research costs for developing polio vaccine may be far more than those spent in the applied research and field trials. Nevertheless, the estimated saving of $ 1 billion per year leaves much leeway for added research or indirect costs, particularly when the vaccine has been used with tremendous success for over 25 years. *Schumacher* [16] estimated that West Germany saved 90 German marks for treatment, care, and rehabilitation for every German mark spent for polio vaccination during 1962–1970.

There is little argument as to the cost-effectiveness of poliomyelitis vaccine. However, the controversy over the relative values of killed versus live polio virus vaccine [17–19] might well be aided by formal benefit-cost or cost-effectiveness analysis.

Measles

Prior to the licensing of the measles virus vaccine in 1963, measles incidence was approximately 4 million cases in the United States each year. The major clinical and public health impact of the disease lay in its complications – 4,000 cases of measles encephalitis and 400–500 deaths annually. In a benefit-cost analysis of measles immunization, the effect of the vaccine on the incidence of measles was determined by projecting the incidence of measles with and without the use of a vaccine [5]. The incidence of measles without a vaccine was based on the incidence of the previous decade projected for changes in the birth rate and epidemicity of measles. Using these epidemiologic data and projections, the net benefits of measles immunization were estimated to be $ 1.3 billion for the period 1963 through 1972 (discounted at a rate of 4%), releasing substantial resources for other uses. Medical savings included 1.4 million hospital days and more than 12 million physician visits. Savings in educational resources included 75 million school days and because the program averted 7,900 cases of mental retardation, it averted substantial expenditures for special schooling. By averting retardation and premature death, it assured that more than 10,300 persons would have an opportunity to lead productive lives measurable at 709,000 years. The benefits due to immunization clearly exceed the costs of immunization. The benefit-cost ratio was approximately 10:1.

Albritton [20] utilized a time series model in estimating the costs and benefits associated with the current measles elimination efforts in the United States conducted under the leadership of the Center for Disease Control. The results indicated that the impact of federally supported measles eradication efforts on measles incidence was statistically significant using the Box-Tiao time series model. *Albritton* concludes that on the basis of economic considerations alone, support of measles eradication as a continuing component of federal public policy is justified. The benefit-cost ratio of 10:1 represents a sizable net return on investments of resources in preventive health care delivered by public health efforts. However, as the incidence of measles declines, marginal reductions become increasingly more costly.

In a benefit-cost study of the measles immunization program in Finland, where all 1-year-old children are immunized, the benefits of measles immunization for the period 1975–1999 were estimated to be 158.6 million marks, while the costs of immunization were estimated to be 40.8 million marks (discounted at a rate of 6%). The benefit-cost ratio was 3.7:1 [21].

Rubella

Several studies have been undertaken to assess the benefits and costs of immunization against rubella. Rubella usually is a mild illness without complications or sequelae. However, rubella infection of women early in pregnancy frequently results in fetal damage; this is particularly noted in epidemic periods. In 1964–1965, an epidemic resulted in an estimated 5,000 therapeutic abortions, 20,000 children who were born with congenital rubella sequelae, and 2,100 excess neonatal deaths associated with the epidemic [22]. Of the 20,000 children born with congenital rubella syndrome, some 8,000 were deaf, 3,500 deaf and blind, and several thousand suffered moderate to severe mental retardation.

The estimated direct costs associated with the 1964–1965 epidemic were $ 1 billion. The major component of the direct care costs was the long-term care associated with congenital rubella syndrome which accounted for about 90% of the direct costs. Special educational care costs, that is, the educational care costs in excess of those for regular education, accounted for 80% of the long-term costs and institutional care for about 16%. Acute care costs for physician and hospital services only contributed 10% of total costs.

Given the US Public Health Service policy to immunize children, age 1–12 years of age, the cumulative cost of immunization would have to exceed $ 1 billion in 1964–1965 dollars before this immunization effort would fail to be beneficial in terms of direct costs for a major epidemic which was expected to occur every 8–10 years.

Elo [23] determined that programs aimed at the elimination of rubella from the community by immunization of children and programs aimed at immunization of girls prior to adolescence are both economically beneficial in Finland.

Programs aimed at the immunization of girls prior to adolescence were economically more beneficial than programs aimed at the immunization of children. However, the analysis did not consider in its calculations the effect that vaccination of children would have in reducing the number of infectious rubella contacts to which pregnant women might be exposed. *Elo* noted that the benefits from measles and rubella immunization are approximately of the same magnitude, but that the benefits of rubella immunization are largely related to the prevention of fetal damage, whereas for measles immunization the benefits are largely the prevention of costs associated with neurologic complications of the disease.

Schoenbaum et al. [24] compared the programs in the United States and the United Kingdom with respect to the immunization of different target populations. The United States emphasizes immunization of children between the ages of 1 and 12 years while the United Kingdom directs its programs towards girls between the ages of 11 and 14 years. In 1972, the direct cost of acute rubella for 1 million persons was estimated to be $ 2.7 million. The costs of congenital rubella syndrome occurring in the offspring of 1 million unprotected females was estimated to be $ 35.9 million. Based on the current rates of infection, *Schoenbaum's* results indicated that rubella immunization has economic benefits at any age but the benefits were greater if the vaccine was offered once at the age of 12 years to females as compared to children ages 2 or 6 years. The benefit-cost ratio for monovalent rubella vaccine given to 1 million females at the age of 12 years was estimated to be 25:1 with 100% immunization of the target population. If the immunization rate was 80%, *Schoenbaum* estimated that the fewest number of children born with congenital rubella syndrome would occur when the vaccine is offered to all children at age 2 years. *Schoenbaum* recognizes that since the ultimate target population for the rubella immunization program is the unborn fetus, the availability and acceptability of abortion are critical factors in assessing the benefits.

Mumps

Since mumps vaccine was licensed in 1967, more than 40 million doses have been distributed in the United States, mostly in combination with measles and rubella vaccines (MMR) [25]. The incidence of mumps has declined from rates of 90–200 per 100,000 population before 1967 to rates of 7–10 per 100,000 in the late 1970s [26]. During this period there has also been an increase in immunization coverage for mumps.

When a model is used following the hypothetical experience of a cohort of 1 million children from age 1 to 30 years of age, comparing the alternatives of receiving mumps vaccination at 1 year and not receiving mumps vaccine, a benefit-cost ratio can be determined [27].

Under the set of assumptions used in the analysis, the use of mumps vaccine for a cohort of 1 million persons would prevent over 74,000 cases of mumps and 3 deaths. A comparison of the benefits of vaccination over a 30-year period (the savings of prevented disease minus the cost of vaccine administration) and the costs of mumps in the absence of a vaccination program, shows that a program which includes mumps vaccine would reduce the costs associated with mumps by over 86% – $ 846,827 versus $ 6,271,764 for each million children, a net benefit of $ 5,424,937. The benefit-cost ratio is thus 7.4:1. A benefit-cost ratio of 1 would be reached when vaccine and administration costs were $ 8.23 per dose. This cost might be easily reached if mumps vaccine were administered as a separate antigen, rather than as part of a measles-mumps-rubella combination. Mumps vaccine use in Austria has been estimated to have a benefit-cost ratio of 3.6:1 (administration costs included and the model cohort followed for 12 years) [28]. In Switzerland, the benefit-cost ratio of mumps vaccine where used in the 2nd year of life is estimated at 2.1:1 (without consideration of indirect costs) [29].

Influenza

The Office of Technology Assessment of the US Congress is conducting a cost-effectiveness analysis of influenza vaccination [*M.A. Riddiough*, personal commun.]. Preliminary results indicate that medical care costs associated with influenza vaccination during 1971–1978 totalled $ 808 million. 150 million people were vaccinated and approximately 13 million years of healthy life added, giving a per vaccination cost of $ 63 per year of healthy life gained, averaging all age groups. However, the cost-effectiveness of influenza vaccination improves with increasing age of the vaccinee, from $ 258 per vaccination for each year of healthy life gained in the under-3-year-old age group to $ 23 per vaccination in the 45- to 64-year-old age group to a positive cost savings in the age group 65 years or older. When vaccination is confined to high-risk individuals (those most susceptible to influenza morbidity and mortality) and one excludes medical care costs, the cost-effectiveness of the procedure is increased.

Schoenbaum et al. [30] analyzed the economic aspects of the 1976–1977 mass immunization against swine influenza, using the Delphi technique to provide estimates on probabilities.

With an epidemic probability of 0.10 and vaccine costs of 50 cents per person, the program was estimated to be economically justifiable if restricted to adults 25 years of age or over and acceptance rates exceeded 50%. Although the consensus of the experts polled in the study was that over 60% of the general population would accept swine flu vaccine, in reality only 6 states attained influenza immunization participation rates of 60% or greater [31].

A benefit-cost analysis of influenza vaccine in Finland considered various population groups as vaccine recipients: the employed labor force, health personnel and medical risk groups [32]. Calculations were made for a 10-year period with two influenza vaccinations given in the 1st year and one annually thereafter. The alternatives of attenuated live vaccine, killed vaccine and drug prevention (Amantadine) were considered. The most profitable preventive measure was use of an attenuated virus vaccine. The net benefits over a 10-year period were estimated to be 100–600 million Finnish marks (the range depending on varying assumptions of vaccine efficacy, 30–60%). The benefit-cost ratio of programs using attenuated vaccine in the employed labor force were calculated as 1.3:1 to 2.5:1. This study does not consider mortality and prevention of mortality in its calculations, thus possibly underestimating the benefits of influenza vaccination.

Summary

The techniques of benefit-risk, benefit-cost and cost-effectiveness analysis have been applied to several vaccines. The analyses are useful to the scientific community, and to public health practitioners and administrators. The studies can also put vaccination programs in an economic perspective for policy makers, legislators, other budget allocators and planners. They permit investigators to focus on crucial variables in decision-making which are in need of better biologic or epidemiologic information and thus help direct research. They promote the use of valuable vaccines in places not currently engaged in vaccination programs (e.g., measles) and discourage the misuse of inappropriate vaccines (e.g., smallpox). It is worth noting that the studies we have reviewed are aimed at public health, not at individual clinical decision-making. Benefits, risks and costs have been viewed from a societal perspective. For decisions on the use of viral vaccines for particular individuals, different conclusions might be reached. Thus, killed poliomyelitis vaccine might be preferable for some persons. Boys receive rubella vaccine less for their immediate benefit than for the benefit of society as a whole and the role these young male recipients play in disease transmission to potentially infected pregnant contacts.

The variables that make a given vaccine cost-beneficial vary with each vaccine. The long-term disabilities of paralytic poliomyelitis and the complications of measles constitute major cost factors for these diseases. Most of the costs of mumps arise from the acute illness. The immediate complications of the acute illness are major cost factors in influenza. For rubella, the considerable disabilities of fetal infection provide the major costs.

Benefit-risk, benefit-cost and cost-effectiveness analyses can be useful in determining target groups or strategies for vaccination programs. The newly developed hepatitis B vaccine might be most economically distributed with the largest public health impact if a benefit-cost analysis were done to determine which target populations should receive it.

References

1 Weinstein, M.; Stason, W.B.: Foundations of cost-effectiveness analysis for health and medical practices. New Engl. J. Med. *296:* 716–721 (1977).

2 Shephard, D.S.; Thompson, M.S.: First principles of cost-effectiveness analysis in health. Publ. Hlth Rep., Wash. *94:* 535–543 (1979).

3 Office of Technology Assessment, U.S. Congress: The implications of cost-effectiveness analysis of medical technology. GPO Stock No. 052-003-00765-7, Washington, D.C., and US Government Printing Office, August 1980.

4 Weisbrod, B.A.: Costs and benefits of medical research: a case study of poliomyelitis. J. pol. Econ. *79:* 527–544 (1971).

5 Witte J.J.; Axnick, N.W.: The benefits from 10 years of measles immunization in the United States. Publ. Hlth Rep., Wash. *90:* 205–207 (1975).

6 Center for Disease Control: Public Health Service Recommendation on Smallpox Vaccination, and vaccination against smallpox in the United States; a reevaluation of the risks and benefits. Morbidity Mortality Weekly Rep. *20:* 339–345 (1971).

7 Dixon, C.W.: Smallpox (Churchill, London 1962).

8 Neff, J.M.; Levine, R.H.; Lane, J.M.; Ager, E.A.; Moore, H.; Rosenstein, B.J.; Millar, J.D.; Henderson, D.A.: Complications of smallpox vaccination, United States, 1963. II. Results obtained by four statewide surveys. Pediatrics, Springfield *39:* 916–923 (1967).

9 Neff, J.M.; Lane, J.M.; Pert, J.H.; Moore, R.; Millar, J.D.; Henderson, D.A.: Complications of smallpox vaccination. I. National survey in the United States, 1963. New Engl. J. Med. *276:* 125–132 (1967).

10 Lane, J.M.; Ruben, E.L.; Neff, J.M.; Millar, J.D.: Complications of smallpox vaccination, 1968. National surveillance in the United States. New Engl. J. Med. *281:* 1201–1208 (1969).

11 Lane, J.M.; Ruben, F.L.; Neff, J.M.; Millar, J.D.: Complications of smallpox vaccination, 1968. II. Results of ten statewide surveys. J. infect. Dis. *122:* 303–309 (1970).

12 Lane, J.M.; Ruben, F.L.; Abrutyn, E.; Millar, J.D.: Deaths attributable to smallpox vaccination, 1959 to 1966, and 1968. J. Am. med. Ass. *212:* 441–444 (1970).

13 Neff, J.M.: The case for abolishing routine childhood smallpox vaccination in the United States. Am. J. Epidem. *93:* 245–247 (1971).

14 Sencer, D.J.; Axnick, N.W.: Cost benefit analysis; In Perkins, Proc. Int. Symp. on Vaccination Against Communicable Diseases, Monaco 1973. Symp. Ser. immunobiol. Standard., vol. 22, pp. 37–46 (Karger, Basel 1973).

15 Fudenberg, H.H.: Fiscal returns of biomedical research. J. invest. Derm. *61:* 321–329 (1973).

16 Schumacher, W.: Discussion of cost-benefit analysis; in Perkins, Proc., Int. Symp. on Vaccination Against Communicable Diseases, Monaco 1973. Symp. Ser. immunobiol. Standard., vol. 22, pp. 47–48 (Karger, Basel 1973).

17 Nightingale, E.O.: Recommendations on a national policy on poliomyelitis vaccination. New Engl. J. Med. *297:* 249–253 (1977).

18 Melnick, J.L.: Advantages and disadvantages of killed and live poliomyelitis vaccines. Bull. Wld Hlth Org. *56:* 21–38 (1978).

19 Assaad, F.: Poliomyelitis vaccination benefits – versus risk. Int. Symp. on Immunization: Benefit Versus Risk Factors, Brussels 1978. Dev. biol. Standard., vol. 43, pp. 141–150 (Karger, Basel 1979).

20 Albritton, R.B.: Cost-benefits of measles eradication: effects of a Federal intervention. Policy Analysis *4:* 1–22 (1978).

21 Ekblom, M.; Elo, O.; Laurinkari, J.; Niemela, P.: Costs and benefits of measles vaccination in Finland. Scand. J. soc. Med. *6:* 111–115 (1978).

22 Sencer, D.J.; Axnick, N.W.: Utilization of cost/benefit analysis in planning prevention programs. Acta med. scand. suppl. 576, pp. 123–128 (1975).

23 Elo, O.: Health and economic consequences of rubella. Acta Univ. tamper., ser. A *60:* 27–47 (1974).

24 Schoenbaum, S.C., Hyde, J.N.; Bartoshesky, L.; Crampton, K.: Benefit-cost analysis of rubella vaccination policy. New Engl. J. Med. *294:* 306–310 (1976).

25 Center for Disease Control: Biologics Surveillance Report No. 79. Annual Summary, July–December 1979.

26 Center for Disease Control: Mumps Surveillance Report. July 1974–December 1976 (issued July 1978). HEW Publ. No. (CDC) 78-8246, 1978.

27 Koplan, J.P.; Preblud, S.R.: A benefit-cost analysis of mumps vaccine (unpublished report, 1981).

28 Wiederman, C.; Ambrosch, F.: Costs and benefits of measles and mumps immunization in Austria. Bull. Wld Hlth Org. *57:* 625–629 (1979).

29 Just, M.: Rentiert die Masern- und/oder Mumps-Impfung für schweizerische Verhältnisse? Schweiz. med. Wschr *108:* 1763–1768 (1978).

30 Schoenbaum, S.C.; McNeil, B.J.; Kavet, J.: The swine-influenza decision. New Engl. J. Med. *295:* 759–765 (1977).

31 Schoenbaum, S.C.: Influenza vaccine – unacceptable or unaccepted. Am. J. publ. Hlth *69:* 219–221 (1979).

32 Elo, O.: Cost-benefit studies of vaccinations in Finland. Int. Symp. on Immunization: Benefits Versus Risk Factors, Brussels 1978. Dev. biol. Standard., vol. 43, pp. 419–428 (Karger, Basel 1979).

Dr. Jeffrey P. Koplan, Office of Program Planning and Evaluation,
Centers for Disease Control, Atlanta, Ga 30333 (USA)

Prog. med. Virol., vol. 28, pp. 192–207 (Karger, Basel 1982)

A Look at the P4 Virus Containment Laboratory

Guido van der Groen, Philip C. Trexler

Institute of Tropical Medicine, Antwerp, Belgium; University of Surrey, Guildford, England

Contents

I. Introduction

In the beginning of October 1976, specimens from a serious outbreak of hemorrhagic fever in Zaïre were sent to the virological laboratory of the Tropical Institute of Medicine, Antwerp.

A new virus was isolated, and termed Ebola [16]. About four hundred persons died during the epidemic and the reported death-to-case ratio varied from 55% in Sudan [3] to 88% in Zaïre [5].

Unknowingly the first Ebola virus isolation was made on an open table without any protection for personnel and environment. This Ebola isolation was confirmed at the same time by the Center for Disease Control, Atlanta, Georgia, USA [13] and the Public Health Laboratory Service Centre for Applied Microbiology and Research, Special Pathogens Reference Laboratory, Porton Down, Salisbury, Wiltshire, England [4], two of the three laboratories at that time having a class 4 or maximum containment laboratory, i.e., a laboratory specially designed to provide maximum containment of the agent as well as personnel protection. In such a laboratory all manipulations must be performed in a gas-tight glovebox or isolator, avoiding at all times direct contact between the infectious agent and the investigator.

Very unfortunately, one of the investigators at the high-security laboratory in England accidently pricked his thumb, while manipulating this virus. It was proven that he had the disease, and he recovered satisfactorily [10].

This incident proves that even the best-equipped laboratory is not able to give complete protection to the investigator.

Why? Because the safety of a system is always determined by the weakest point in the system. Experiences like this and others have shown that the gloves are the most puncture-prone part of the barrier separating virus and investigator.

As long as the risk of thin-wall gloves is accepted, why use expensive isolators with thick, rigid walls? Cannot the same work be performed just as safely in a less expensive flexible-film type of isolator?

What is the *minimum* infrastructure necessary for a laboratory to handle safely class 4 viruses?

What criteria determine the necessity to use a P4 containment laboratory when working with a particular virus?

Problems of this kind are the subject of this paper.

II. What Are Class 4 Viruses?

How dangerous is a virus and what is the risk for an investigator handling it in a laboratory? It is very difficult to answer this question; the hazard is a function of many factors:

- Virulence of the virus for man and animals.
- Capability for causing serious epidemic diseases in humans or animals.
- Occurrence of fatal human laboratory infections.
- Capacity of the virus to produce aerosol infections.
- The frequency of overt laboratory infections and the consequences for the victims.
- Nature and kind of study in which the virus is being used. Aerosol studies, inoculation of animals and arthropod vectors markedly increase the hazard, whereas in vitro experiments decrease the hazard.

Table I. Class 4 viruses on the basis of hazard for humans

Everglades
Mucambo
Venezuelan equine encephalitis
Yellow fever, wild strain
Absettarov
Hanzalova
Hypr
Kumlinge
Kyasanur Forest disease
Omsk hemorrhagic fever
Russian spring-summer encephalitis
Congo-Crimean hemorrhagic fever
Korean hemorrhagic fever
Ebola
Marburg
Junin
Lassa
Machupo
Rift Valley Fever

- The public health risk accompanying animal infection or transmission experiments in areas where the natural host and vector are present.
- Infection prevalence in natural human and/or animal population.
- Availability of specific vaccines.

Viruses were classified on basis of hazard into four groups, class 1 the non-dangerous, and class 4 the most dangerous group [1, 2, 7].

Generally, viruses belong to the class 4 virus group when

- they cause serious epidemic disease;
- they cause disease which may have a high mortality rate in man;
- they cause disease for which no treatment is available;
- they cause disease which can be transmitted from person to person by aerosols;
- there is no vaccine available;
- they are in close antigenic relationship with a known level 4 agent and sufficient laboratory experience is lacking.

The viruses actually belonging to hazard group 4 on the basis of hazard for *humans* are summarized in table I.

The viruses Everglades, Mucambo, Venezuelan equine encephalitis and yellow fever, wild strain, belong to class 4 only when personnel handling these viruses are not vaccinated.

Table II. Geographical distribution of diseases caused by class 4 viruses, confirmed by isolation of the virus

Russian spring-summer encephalitis virus: USSR
Kyasanur Forest disease virus: India (Mysore state)
Omsk hemorrhagic fever virus: USSR (Siberia)
Congo-Crimean hemorrhagic fever virus: Asia, USSR, Bulgaria;
Southeast Asia: Pakistan, Iraq;
West Africa: Senegal;
Central Africa: Uganda, Zaïre, Central African Empire;
East Africa: Kenya, Ethiopia
Marburg virus: West Germany, Yugoslavia (laboratory accidents),
South Africa (Johannesburg), Rhodesia, Kenya (Nairobi)
Ebola virus: Northern Zaïre (Yambuku area), Northwestern Zaïre (Tandala-Gemena area),
Southern Sudan (Nzara – Maridi district)
Junin virus: Argentina (Buenos Aires)
Machupo virus: Bolivia (Beni region)
Lassa virus: West Africa, Nigeria, Liberia, Sierra Leone
Korean hemorrhagic fever virus: USSR, Czechoslovakia, Yugoslavia, Japan, Korea

The viruses Absettarov, Hanzalova and Kumlinge are placed in level 4, based on the close antigenic relationship with Russian Spring-Summer encephalitis virus and insufficient laboratory experience.

Rift Valley Fever virus will best be treated as a class 4 dangerous virus based on the recent epidemics in Egypt [15], especially when handled by persons who are not vaccinated.

It is clear that the list of viruses in table I is not a definitive one and will change continuously, on the basis of new epidemiological information and laboratory experiences. Therefore, each investigator must use scientific judgment to decide whether a virus should be handled in a P4 laboratory or not.

Which are the criteria that can help him to make the decision? His decision will be based mainly on epidemiological information which in the best case will include the following points:

- Exact geographical situation of the area where the sample came from or where the patient stayed 3 weeks before onset of disease. Suspect areas for class 4 agents are summarized in table II.
- Exposure to a rural environment.
- Direct contact with ill individuals coming from such areas.
- Contact with animals, rodents and their excreta.
- Contact with insects (ticks).
- Occupation of the patients: medical and nursing staff from country hospitals, missionary workers, laboratory workers handling dangerous viruses.

- Illness despite adequate malarial prophylaxis or when malaria is excluded by negative blood smears.
- Sample from an area where a high number of deaths occurred without any further information.
- Suspect clinical picture with the following symptoms: headache, sore throat, abdominal pain, myalgia, arthralgia, bleeding, diarrhea, melena, oral throat lesions, vomiting, maculopapular rash, conjunctivitis, cough.
- Biological sample from an area suspect for class 4 agents without any accompanying information.

It remains difficult to decide whether or not a sample must be treated in a maximum security laboratory. *Emond* et al. [9] received over a 2-year period 46 samples from patients suspected of having an African-derived hemorrhagic fever; only 2 had such infections.

III. Methods of Physical Containment

The objective of physical containment is to reduce the exposure of laboratory workers (primary containment) and the environment (secondary containment) to microorganisms and biological materials which may or do pose a health hazard.

Primary containment is accomplished by the use of special laboratory practices and equipment security.

Secondary containment includes facility security measures, such as architectural and construction features, waste sterilization and special ventilation systems.

Experiments involving class 4 viruses should be performed in a high containment laboratory (HCL) or maximum security laboratory (MSL), where both the primary and secondary containment levels are maximum.

A. Primary Containment

The maximum level of primary containment is provided by the use of class 3 cabinet systems for all procedures and operations involving the experimental material.

A class 3 microbiological safety cabinet [6] is defined as a safety cabinet by which the operator is separated from the work by a physical barrier. It is kept under negative pressure, and the escape of any airborne particle is prevented by an exhaust system. The negative pressure prevents escape of contaminated air through accidental leaks in the cabinet.

Operations within the cabinet are conducted through attached rubber gloves. The extract air is either filtered through two HEPA (High Efficiency Particulate Air) filters, or incinerated following filtration through a single HEPA filter.

Class 3 cabinets or isolators incorporate equipment which allows the safe introduction and removal of materials. This equipment can be: a double-door autoclave, chemical dunk tank, double-door fumigation chamber and a heat-sealable disposable plastic bag system [20]. Two types of class 3 safety cabinets or isolators have been developed: isolators with rigid walls and isolators with flexible walls.

1. Isolators with Rigid Walls

These are constructed in stainless steel with transparent Plexiglas window and gloves of flexible material. This type of isolator is currently used in the existing maximum security laboratories of the Center for Disease Control (CDC), Atlanta, Ga., the US Army Medical Research Institute for Infectious Diseases, Frederick, Md., the Public Health Laboratory Service Centre for Applied Microbiology and Research, Porton Down, England, and the National Institute for Virology, Johannesburg, South Africa.

These isolators usually are connected in line and operated with an internal pressure significantly negative to that in the laboratory. Double-doored autoclaves and chemical dunk tanks are employed to pass materials into and out of the cabinet line. The chemical dunk tank is provided for the safe removal from the class 3 cabinet of material and equipment that cannot be heat-sterilized, for example, virus harvests. Most of the time the tank is filled with sodium hypochlorite solution to surface-sterilize the items brought out.

2. Isolators with Flexible Walls

A negative-pressure flexible-film laboratory isolator was recently developed [20] and successfully used in the maximum security laboratory of the Tropical Institute of Medicine, Antwerp, Belgium (fig. 1). The isolator system consists of an envelope, made of polyvinylchloride film, totally enclosing the working space. The dimensions are such that one person can reach every corner of the isolator. The supporting frame is outside the envelope. Manipulations within the isolator are accomplished through two pairs of gloves situated at each longitudinal side. The work in the sleeves is facilitated by a special elbow support which is mounted inside the isolator over the complete length and on both sides.

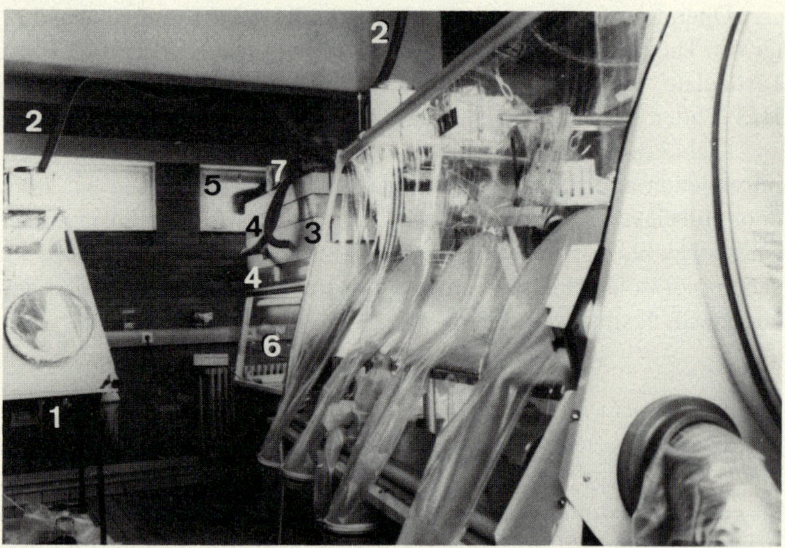

Fig. 1. The maximum security laboratory of Antwerp contains two plastic isolators with flexible wall, each with its own motor to create negative pressure inside. The HEPA-filtered *(1)* exhaust air of the isolators is evacuated through a tube *(2)* to a plenum *(3)* and HEPA-filtered *(4)* for a second time before leaving the building *(5)*. The exhaust air of the MSL itself is evacuated through a biosafety cabinet type 1 *(6)* and is HEPA-filtered twice *(4)*. The MSL is maintained at a negative pressure in relation to the rest of the building by means of a central fan *(7)*.

To assure stability of small bottles and vials, the bottom of the isolator is covered with a heavy metal Inox plate. The isolator has a supply port on both ends. The larger supply port is attached to the incubator end of the isolator.

Plastic film cones at several sites of the isolator allow the entry of electric wires for a power supply to instruments. The isolator has two long small plastic sleeves on both sides which can be placed within a small refrigerator located on a trolley under the isolator, to provide cold storage without the necessity of removing materials from the isolator.

Tissue cultures kept *inside* can be read through the transparent plastic with the inverted microscope standing *outside* the isolator.

Sterile air is introduced into the isolator through two HEPA filters, and the outgoing air passes through one HEPA filter. The air pressure inside is kept continuously at a negative pressure of a 3- to 5-mm water column by means of a motor blower.

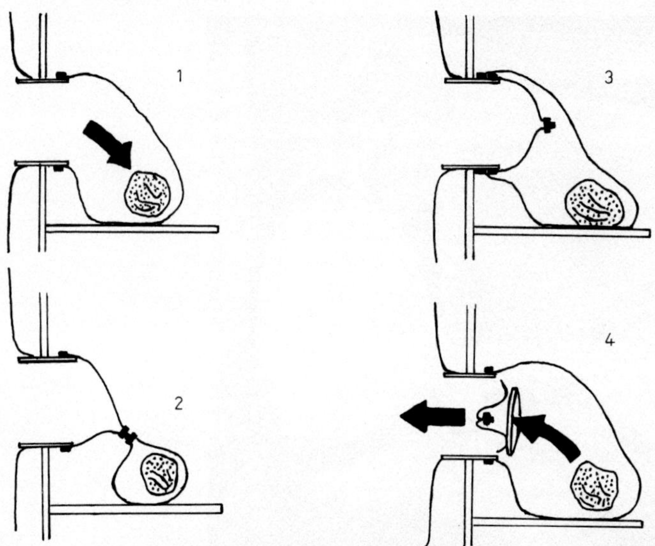

Fig. 2. Materials to be removed are placed into the supply port bag *(1)*, which is gathered to form a neck and tied off with two plastic bag ties. The bag is sprayed with hypochlorite and cut between the two ties *(2)* with a hot-wire sealing device, leaving the supply port closed and the outcoming material in a sealed bag. To introduce material, a new bag is slid over the old bag *(3)* and secured to the supply port by the second rubber band. The old bag is pulled into the isolator *(4)*. The materials are now inside the isolator, the new plastic bag closing the supply port.

A system of a tightly fitting polyethylene bag, kept in place by a rubber band on the outside ring of the port, is used to pass materials into and out of the isolator as shown in figure 2, without breaking the physical barrier.

3. One-Piece Plastic Positive-Pressure Suit

Both rigid and flexible-film types of isolators have in common that work performed inside is very time-consuming and the numbers of manipulations are restricted.

In an effort to increase the variety in experiments inside a P4 lab, CDC, Fort Detrick, and very recently the Institute of Virology, Johannesburg, completed new maximum containment laboratories in which the steel cabinet line is replaced by a plastic one-piece suit supplied with air from a source remote from the laboratory (fig. 3). In contrast to working in the isolator, the investigator is now free to move in the laboratory and has access to much larger specialized equipment. The suit is kept under positive pressure to prevent contamination by air from the laboratory entering through leaks in the suit.

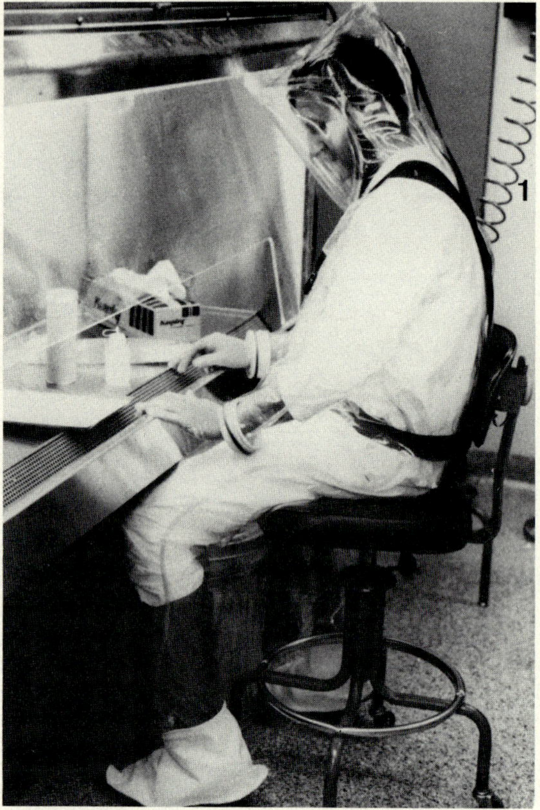

Fig. 3. Plastic one-piece suit supplied with air *(1)* from a source remote from the laboratory. The investigator is now free to move in the laboratory, disconnecting the air supply at the place he is leaving, and connecting his suit again to the air supply at the place of arrival.

B. Secondary Containment

To obtain maximum containment, a class 3 safety cabinet must be used within a maximum containment laboratory. A maximum containment laboratory is by definition designated as a Physical barrier class 4, abbreviated as P4 laboratory [8]. Conditions to obtain a P4 level containment laboratory are the following [1, 2, 17]:

- The maximum containment laboratory is either a separate building or a clearly demarcated and isolated laboratory within the building.
- The laboratory has a separate air exhaust and negative pressure with respect to other areas of the building.
- Exhaust air is decontaminated by filtration through two HEPA filters or incinerated following filtration through a single HEPA filter.
- The exhaust air from the class 3 cabinets is discharged to the outdoors through the laboratory exhaust air system.
- Access to the laboratory is restricted to a limited number of individuals under specific control.
- Entry room, change room and shower facility are provided for personnel entry and exit.
- Protective clothing is worn and it is decontaminated before being removed from the laboratory area.
- Sterilization of contaminated material must be performed in the P4 laboratory itself.
- Equipment must be provided for sterilization of material passing out of the class 3 cabinet without having any direct contact outside the cabinet. A double-door autoclave can satisfy this requirement if one door is directly accessible from inside the isolator and the other door only accessible from outside.

Doors must be controlled in such a way that the outer door can be opened only after completion of a sterilization cycle. This system is used in all existing maximum security laboratories, except in the high security laboratory (HSL) of Antwerp, where contaminated material is transferred in a heat-sealed plastic bag from the isolator to a one-door autoclave standing in the P4 facility [20].

- A pass-through tank or fumigation chamber is provided for the safe removal from the maximum containment laboratory of material that cannot be heat-sterilized.
- All liquid effluents from the maximum containment laboratory are collected and sterilized before disposal.

If a one-piece positive-pressure suit is used, entrance to the P4 laboratory must be through an air lock fitted with air-tight doors. A chemical shower is used to decontaminate the surface of the suits before exit of personnel. Suit air supply is provided with alarms and emergency back-up tank air.

Three suit P4 laboratories are in use; the first one is at CDC, Atlanta, Ga., the second at the Institute of Virology, Johannesburg – and a third one at Fort Detrick. Another unit is under construction in West Germany.

IV. Discussion

The quality of a P4 laboratory can be evaluated in three ways:

- by the efficiency of the containment;
- by the ease with which work can be performed and the variety of manipulations possible;
- by the overall cost of construction and operation.

It is difficult to evaluate the safety of a laboratory. Safety is mainly determined by the quality of laboratory practice and discipline of personnel. Manipulations must be performed with utmost care, regardless of the provision of physical containment. There should be an appropriate manual of MSL practice, and appropriate training should be given.

The number of individuals working in a P4 laboratory must be limited, to facilitate enforcement of safe and correct laboratory practice. The personnel must have good self-control and must be honest in reporting immediately the smallest error made during manipulations within the P4 laboratory. Good laboratory practice necessitates a minimum of spills and no aerosol formation of infectious virus inside the class 3 cabinet.

A second element of safety is the effectiveness of containment of the class 3 cabinet itself. This is determined by the quality of the absolute HEPA filters used and the detection of leaks in the construction. Leaks can be detected by applying positive pressure to the isolator.

The isolator can be filled with halogen gases (e.g., Freon) and tiny leaks can then be detected by a sensitive halogen detector [6]. Also, a bubble test can be carried out by applying a 2.5% solution of soft soap in distilled water to all welds [6]. A sodium chloride aerosol or dioctyl phthalate smoke generator should be operated inside the cabinet and leakage from the cabinet detected by means of a portable sodium flame detector unit or a photometer [6]. The integrity of the HEPA filters is tested by the latter method.

To test under normal operational conditions, bacteriophage aerosols are produced inside the isolator and appropriate culture media put outside the isolator to detect any escaped organisms [12]. It is very important to maintain continuous control of the negative pressure inside the isolator. As long as the negative pressure is maintained, small leaks may occur without too much danger.

Are rigid-wall types of isolators safer than flexible-wall types? Even rigid-wall types of isolators need flexible material for the gloves; otherwise the isolator cannot be used. Experience has shown that the gloves are the most puncture-prone part of the barrier and are thus the weakest point of the system [10].

As long as the risk of the thin-wall gloves is accepted, it is unreasonable to invest in very expensive rigid cabinets, when the same work can be performed in a far less expensive flexible-wall type of isolator.

The plastic film used in flexible-wall type of isolators is sufficiently strong to maintain integrity during normal usage. In one author's laboratory the plastic envelope was intact after 1 year of continuous use.

Rigid walls make it difficult to change the construction of the isolator, whereas alterations of a flexible wall are much easier to perform. In Antwerp, the envelope of the isolator was replaced after 1 year of operation without any difficulties. A small entry port with sleeve was installed. Extensive alterations can be made to a flexible film envelope on site. The flexibility of the cabinet walls makes it very easy to check on the existence of negative pressure inside, because of the resulting concave configuration.

In both types of isolators the movements of the investigator are limited. However, the mobility and flexibility of the cabinet walls makes it easier for the operator to work in a flexible isolator than in a rigid-type cabinet.

We must be aware of the fact that work in an isolator is about 10 times more time-consuming than in an ordinary class 2 [6] cabinet. It should be clear now that the costs of constructing and maintaining such P4 laboratories are extremely high. They become even greater when we realize that many manipulations are not possible in P4 laboratories provided with class 3 cabinet lines. For example, high-speed centrifugation, molecular biology techniques (such as liquid scintillation counting, column chromatography, electrophoresis, etc.) are very difficult or even impossible to perform in a classical cabinet line.

Therefore, the P4 laboratory with a plastic one-piece suit under positive pressure was a breakthrough in the P4 containment world. In contrast with the work in a cabinet line, the investigator is now free to move in the laboratory and has free access to a much greater variety of equipment. Another advantage of this system is that after fumigation of the laboratory this can be used as a P3 facility, where the investigator is not wearing a positive-pressure suit. All the equipment inside can still be used. P4 workers who have the choice prefer to work in the suit laboratory rather than in the cabinet line, mainly because the manipulations are easier to perform.

According to the experiences of CDC workers [14], the suit laboratory is less expensive in construction and maintenance than the cabinet line P4 facility. However, the costs remain very high, so that only a few centers in the world can afford this kind of laboratory. Isolators with half-suits, similar to those on bed isolators [12], may provide a practical way of increasing the amount of equipment that can be used within an isolator without requiring a one-piece-suit facility.

Several class 4 viruses are circulating in Africa. Some African countries should have laboratory facilities where rapid and safe diagnosis of these viruses is possible, avoiding the delicate problem of transport of well-cooled and preserved biological samples to laboratories overseas. Naturally, most of the African countries, with the exception of South Africa, do not have the necessary resources to construct and maintain a P4 laboratory like the ones used at the CDC and the PHS Special Pathogens Laboratory of England. On the other hand, it is the authors' opinion that safe handling of class 4 viruses can be performed with a much less expensive containment system, such as the one used in Antwerp [20]. In this laboratory a less expensive type of flexible-wall isolator was used. Lassa, Ebola, Marburg, Rift Valley Fever, and Congo-Crimean viruses have been contained in this system for almost 2 years.

The flexible-wall isolator allowed the performance of all basic virologic techniques necessary for rapid virus diagnosis. Inoculation of cell cultures, incubation, reading cytopathogenic effects by inversion light microscopy, plaque assay, neutralization tests, ELISA tests, virus stock preparations and titrations, inactivation experiments with viruses by the virucidal floating technique [18], preparation of grids for electron microscopy, and indirect immunofluorescence test are currently performed in this type of isolator. A safe procedure was developed to transfer the slides with noninactivated virus for immunofluorescence reading outside the isolator [20].

A second flexible-wall isolator was used to contain animals. Only baby mice with mother were used, because most of the class 4 viruses can grow in the brain of baby mice [19]. The cost-saving in the construction and maintenance of the MSL of Antwerp was mainly achieved by

- the use of two less expensive commercially available prototype (Vickers Medical, England) negative-pressure plastic film isolators;
- the small dimensions of the MSL itself with a total volume of $\pm$ 100 m^3, which facilitates the maintenance of the containment;
- the simple but efficient use of the fan of a type 1 safety cabinet to maintain the MSL constantly at a negative pressure;
- the use of a safe system to transfer contaminated material from inside the isolator to a single-door autoclave;
- the absence of expensive collection and treatment systems for liquid wastes.

The elimination of systems for liquid wastes is achieved by the absence of any water-duct system for handwashing, toilet and shower room. Toilets are outside the P4; the water used for handwashing is collected in a container, which is autoclaved before leaving the laboratory. No shower room was provided, because it was technically very difficult to construct the collection and treatment systems for the shower liquid waste. In addition, the authors are doubtful about the efficacy of a shower for containing a virus inside a P4 facility. Staff exposed to an aerosol of foot and mouth disease virus caused infection among animals, despite having taken a shower bath upon leaving the contaminated area. Huge quantities of virus were still detected in nasal scrapings of the exposed people [7].

It must be remembered that work in a P4 laboratory is always more expensive and causes much more discomfort for the investigator. The worst thing for a P4 laboratory worker is the realization that the same manipulations as those performed inside a P4 can be performed much more easily and quickly in an ordinary virology laboratory. For this reason it is necessary to develop techniques which allow, as much as possible, safe handling of class 4 agents outside a P4 facility.

In the laboratory of one of the authors (in Antwerp), inactivation studies are performed with class 4 agents, in order to obtain noninfectious class 4 agents which are still immunologically reactive. Formaldehyde-inactivated Lassa, Rift Valley Fever and Ebola viruses react with antibodies in immune electron microscopy; this enables a rapid and safe diagnosis to be made. Slides containing only inactivated virus for indirect immunofluorescence testing allow the determination of class 4 virus antibodies in an ordinary laboratory, even in the field [21, 22].

Rapid and simple tests must be developed for antigen detection in the field. A nice example is the passive reverse agglutination test for Lassa antigen determination developed by *Goldwasser* and coworkers [11]. Chicken erythrocytes are coated with specific antibodies against Lassa virus. A drop of the suspension is mixed with the antigen. After 30 min the presence of virus can be detected by reading the agglutination. This test can also be performed with inactivated antigen, permitting an inhibition test for antibodies which can be used in the field for rapid diagnosis.

A simplified P4 facility such as the one used in Antwerp can be sufficient to confirm the results obtained in the field by performing the conventional virologic techniques. It would be best to have such a laboratory in the country where the virus is endemic.

More fundamental research on class 4 viruses should be limited to only a few highly equipped P4 facilities, possessing the positive-pressurized one-piece plastic suit system and the necessary infrastructure for safe handling of larger animals.

Virologists must remain vigilant because, sooner or later, known and still-unknown class 4 viruses will enter their laboratories, although these laboratories will not be of the P4 type. It is obvious that the best containment will always be achieved by those with the best laboratory practice and discipline.

Summary

Class 4 viruses like Ebola, Marburg, Lassa, Congo-Crimean and Rift Valley Fever viruses have been successfully contained during a 2-year period in an inexpensive P4 facility using a negative-pressure flexible-film isolator. All techniques necessary for rapid virus diagnosis of these viruses can be performed in this type of isolator.

More fundamental research on these dangerous viruses is performed preferably in a P4 laboratory where the investigator is protected by a one-piece positive-pressurized plastic suit. The suit system is cost-saving, and access is given to a much greater variety of equipment and manipulations than with a rigid-wall safety cabinet system for primary containment.

References

1 Scherer, W.F.; Eddy, G.; Monath, T.P.; Walton, T.: Survey and recommendation reports of the subcommittee on arbovirus laboratory safety. Arthropod-borne virus information exchange. Special Edition, January, 1978.

2 Scherer, W.F.; Eddy, G.; Monath, T.P.; Walton, T.: Recommendations concerning arboviruses and certain other viruses of vertebrates. Arthropod-borne virus information exchange. Special Edition, January, 1979.

3 Babiker Mohd El Tahir: The haemorrhagic fever outbreak in Maridi, Western Equatoria, Southern Sudan; in Pattyn, Ebola virus haemorrhagic fever, pp. 125–127, (Elsevier/North Holland, Amsterdam 1978).

4 Bowen, E.T.W.; Platt, G.S.; Lloyd, G.; Baskerville, A.; Harris, W.J.; Vella, E.E.: Viral haemorrhagic fever in Southern and Northern Zaïre. Preliminary studies on the aetiologic agent. Lancet *i:* 571–573 (1977).

5 Breman, J.G.; Piot, P.; Johnson, K.M.; White, M.K.; Mbuyi, M.; Sureau, P.; Heymann, D.L.; van Nieuwenhove, S.; McCormick, J.B.; Ruppol, J.P.; Kintoki, V.; Isaācson, M.; van der Groen, G.; Webb, P.A.; Ngvete, K.: The epidemiology of Ebola haemorrhagic fever in Zaïre; in Pattyn, Ebola virus haemorrhagic fever, pp. 103–121 (Elsevier/North Holland, Amsterdam 1978).

6 British Standards Institution: Specification for microbiological safety cabinets (1979).

7 Bruce, W.: Personal commun.

8 Department of Health, Education and Welfare, National Institutes of Health: Federal Register *41:* 27902–27943 (1976).

9 Emond, R.T.D.; Smith, H.; Welsby, P.D.: Assessment of patients with suspected viral haemorrhagic fever. Br. med. J. *i:* 966–967 (1978).

10 Emond, R.T.D.: Isolation, monitoring and treatment of a case of Ebola virus infection; in Pattyn, Ebola virus haemorrhagic fever, pp. 27–32 (Elsevier/North Holland, Amsterdam 1978).

11 Goldwasser, R.A.; Elliott, L.H.; Johnson, K.M.: Preparation and use of erythrocyte-globulin conjugates to Lassa virus in reversed passive hemagglutination inhibition. J. clin. Microbiol. (in press, 1982).

12 Hutchinson, J.G.P.; Gray, J.; Flewett, T.H.; Emond, R.T.D.; Evans, B.; Trexler, P.C.: The safety of the Trexler isolator as judged by some physical and biological criteria: a report of experimental work at two centres. J. Hyg., Camb. *81:* 311–319 (1978).

13 Johnson, K.M.; Webb, P.A.; Lange, V.E.; Murphy, F.A.: Isolation and partial characterization of a new virus causing acute haemorrhagic fever in Zaïre. Lancet *i:* 569–571 (1977).

14 Johnson, K.M.: Personal commun.

15 Laughlin, L.W.; Meegan, J.M.; Strausbaugh, L.J.; Morens, D.M.; Watten, R.H.: Epidemic Rift Valley fever in Egypt: observations of the spectrum of human illness. Trans. R. Soc. trop. Med. Hyg. *73:* 630–633 (1979).

16 Pattyn, S.R.; van der Groen, G.; Piot, P.; Courteille, G.: Isolation of Marburg-like virus from a case of haemorrhagic fever in Zaïre. Lancet *i:* 573–574 (1977).

17 US Department of Health, Education and Welfare, Public Health Service, Center for Disease Control, Office of Biosafety, (Atlanta; Ga.): Classification of etiologic agents on the basis of hazard (1976).

18 van der Groen, G.; El Mekki, A.A.; Pattyn, S.R.: A new method for virucide testing. J. vir. Methods *1:* 27–31 (1980).

19 van der Groen, G.; Jacob, W.; Pattyn, S.R.: Ebola virus virulence for newborn mice. J. med. Virol. *4:* 239–240 (1979).

20 van der Groen, G.; Trexler, P.C.; Pattyn, S.R.: Negative-pressure flexible film isolator for work with class IV viruses in a maximum security laboratory. J. Inf. *2:* 165–170 (1980).

21 van der Groen, G.; Pattyn, S.R.: Measurement of antibodies to Ebola virus in human sera from N.W. Zaïre. Annls Soc. belge Méd. Trop. *59:* 87–92 (1979).

22 van der Groen, G.; Johnson, K.M.; Webb, F.A.; Wulff, H.; Lange, J.: Results of Ebola antibody surveys in various population groups; in Pattyn, Ebola virus haemorrhagic fever; pp. 203–205 (Elsevier/North Holland, Amsterdam 1978).

Dr. G. van der Groen, Institute of Tropical Medicine 'Prince Leopold',
Nationalestraat 155, B-2000 Antwerp (Belgium)

Prog. med. Virol., vol. 28, pp. 208–221 (Karger, Basel 1982)

Taxonomy and Nomenclature of Viruses, 1982

Joseph L. Melnick

World Health Organization Collaborating Centre for Virus Reference and Research, Department of Virology and Epidemiology, Baylor College of Medicine, Houston, Tex., USA

Contents

Introduction

In the past year a number of important developments have taken place in viral taxonomy. Much progress is reflected in the proposals brought before the International Committee on Taxonomy of Viruses at its meetings during the Fifth International Congress for Virology, held in Strasbourg, August 2–7, 1981. Not only have new families and genera been proposed, but also beginnings are being made on definition of species within some genera of viruses. New developments of special interest to medical virology are depicted in the tables and discussed in the text of this chapter. The details of the ICTV decisions will be published as the Fourth Report of the International Committee on Taxonomy of Viruses.

Table I. DNA-containing viruses with cubic symmetry and naked nucleocapsid

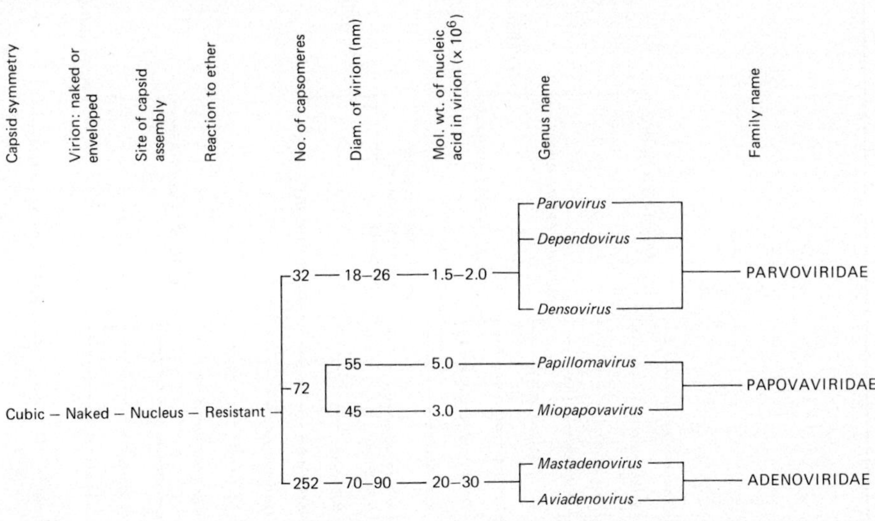

Viruses with a DNA Genome

In table I are shown vertebrate viruses having a DNA genome, cubic capsid symmetry, and a naked nucleocapsid (Parvoviridae, Papovaviridae, and Adenoviridae).

A previously unresolved problem was the naming of the genus within Parvoviridae that includes the helper-virus-dependent viruses. Since not only adenovirus but also herpesvirus is able to serve as helper for these agents – previously known as 'adeno-associated viruses' or 'adeno-satellite viruses' – this genus has now been named *Dependovirus*. It is also latinized in conformity with accepted viral nomenclature.

In the family Papovaviridae, the name *Papillomavirus* has been generally accepted as the designation for the genus that includes wart viruses of a number of species. Less satisfactory has been the naming of the other genus in this family, which has as its type species the polyoma virus of mice and includes SV40 virus of the rhesus monkey, the JC and BK viruses of man, and several other agents. The first two ICTV Study Groups for this family recommended that the genus be named *'Papovavirus B'*, but the ICTV Executive Committee preferred *'Polyomavirus'*. However, members of the Third

Table II. DNA-containing viruses with envelopes or complex coats

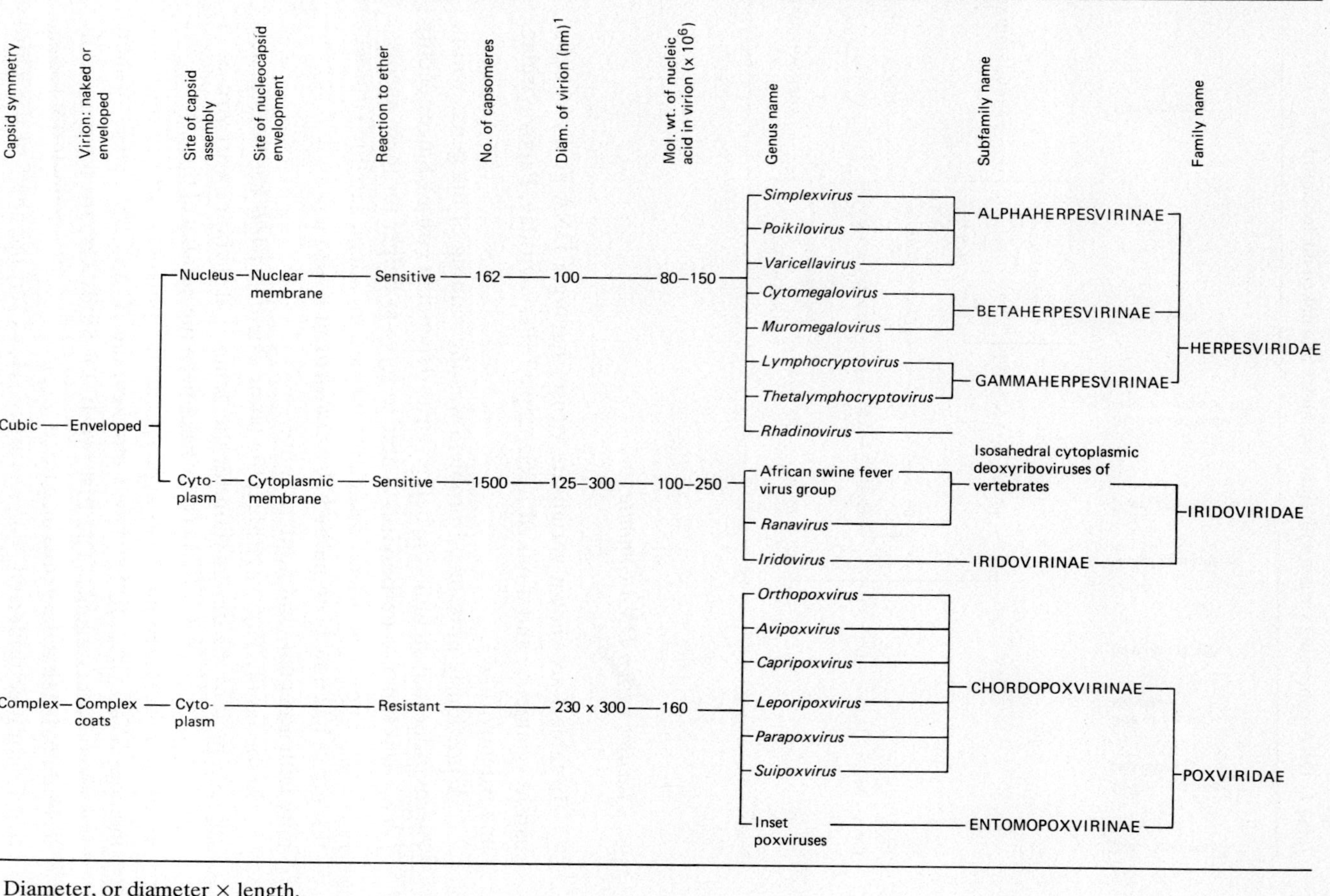

Capsid symmetry	Virion: naked or enveloped	Site of capsid assembly	Site of nucleocapsid envelopment	Reaction to ether	No. of capsomeres	Diam. of virion (nm)[1]	Mol. wt. of nucleic acid in virion ($\times 10^6$)	Genus name	Subfamily name	Family name
Cubic	Enveloped	Nucleus	Nuclear membrane	Sensitive	162	100	80–150	*Simplexvirus*	ALPHAHERPESVIRINAE	HERPESVIRIDAE
								Poikilovirus		
								Varicellavirus		
								Cytomegalovirus	BETAHERPESVIRINAE	
								Muromegalovirus		
								Lymphocryptovirus	GAMMAHERPESVIRINAE	
								Thetalymphocryptovirus		
								Rhadinovirus		
		Cytoplasm	Cytoplasmic membrane	Sensitive	1500	125–300	100–250	African swine fever virus group	Isosahedral cytoplasmic deoxyriboviruses of vertebrates	IRIDOVIRIDAE
								Ranavirus		
								Iridovirus	IRIDOVIRINAE	
Complex	Complex coats	Cytoplasm		Resistant		230 x 300	160	*Orthopoxvirus*	CHORDOPOXVIRINAE	POXVIRIDAE
								Avipoxvirus		
								Capripoxvirus		
								Leporipoxvirus		
								Parapoxvirus		
								Suipoxvirus		
								Inset poxviruses	ENTOMOPOXVIRINAE	

[1] Diameter, or diameter × length.

ICTV Papovaviridae Study Group, who had responsibility for taxonomy of this family through 1981, objected to calling human members of this genus 'human polyomaviruses', a name that would mean 'viruses producing many types of human tumors', when there is no evidence that these viruses are associated with human neoplasms. The Study Group recommended the name *Miopapovavirus* (from the Greek prefix, *mio*, 'smaller', plus *papovavirus*), since the members of this genus are indeed the smaller papovaviruses. The virion diameter is about 45 nm and their DNA molecular weight is 3×10^6, distinct from the 55-nm diameter of the papillomaviruses and their DNA molecular weight of about 5×10^6.

Among the DNA-containing vertebrate viruses with envelopes or complex coats (table II), Herpesviridae had presented considerable problems of taxonomy that are now being resolved [*Roizman* et al., 1982]. From the many possible approaches for grouping herpesviruses, the Herpesviridae Study Group has chosen biological properties as the most practicable to serve as the dominant criteria for classification into subfamilies. Thus host range, duration of reproductive cycle, cytopathology, and characteristics of latent infection form the chief bases for establishment of the subfamilies, Alphaherpesvirinae, Betaherpesvirinae, and Gammaherpesvirinae. The proposals for classification of herpesviruses into genera seem to reflect true phylogenetic relationships; the generic groupings are based on genome structure, serologic relationships, and nucleotide sequence homology.

Alphaherpesvirinae are rapidly growing, highly cytolytic viruses, characterized by a variable host range and a short reproductive cycle with rapid spread of infection in cell culture, resulting in mass destruction of susceptible cells. The establishment of carrier cultures of susceptible cells harboring non-defective genomes is difficult to accomplish. Latent infections are seen frequently, but not exclusively, in ganglia. Three genera have been proposed for viruses in this subfamily. They are: *Simplexvirus,* to include herpes simplex viruses and similar agents; *Poikilovirus,* to include pseudorabies-like viruses; and *Varicellavirus,* thus far including only the varicella-zoster virus.

Betaherpesvirinae (slow-growing, cytomegalic viruses) are characterized by a narrow host range in vivo, a relatively long reproductive cycle, and slowly progressing lytic foci in cell culture. Infected cells frequently become enlarged (cytomegalia) both in vitro and in vivo. Inclusions containing DNA are frequently present in both nuclei and cytoplasm. Carrier cultures are easily established. Latent infections may be established in secretory glands, lymphoreticular cells, kidneys and other tissues. Genera within Betaherpesvirinae include *Cytomegalovirus,* containing the human cytomegaloviruses, and *Muromegalovirus,* cytomegaloviruses of mice.

For the Gammaherpesvirinae, the host range in vivo is usually limited to the same family or order as the host that it naturally infects. In vitro, all members of Gammaherpesvirinae replicate in lymphoblastoid cells; some also cause lytic infections in some types of epithelioid and fibroblastoid cells. Viruses in this group are specific for either B or T lymphocytes. In the lymphocyte, infection is frequently arrested either at a pre-lytic stage, with persistence and minimum expression of the viral genome, or at a lytic stage, causing cell death without production of complete virions. The reproductive cycle of these agents is long, as is their cytopathology. Latent infection is frequently demonstrated in lymphoid tissue. The genera proposed for this subfamily are: *Lymphocryptovirus,* including Epstein-Barr virus and agents with similar properties, as well as herpesviruses of baboon and of chimpanzee; *Thetalymphocryptovirus,* Marek's disease herpesvirus and related agents; and *Rhadinovirus,* to include herpesvirus ateles, herpesvirus saimiri, and similar viruses.

The family Iridoviridae, whose best-known members are insect viruses, also includes important viruses of vertebrates: African swine fever virus and a number of viruses of frogs and fish. No human iridovirus is known. Vertebrate iridoviruses are enveloped; iridoviruses that infect insects contain a lipid fraction in the virion as an integral part of the icosahedral shell, but do not have envelopes as such.

Viruses with an RNA Genome

As indicated in table III, there have been no changes in major groupings of the RNA-containing vertebrate viruses that have cubic capsid symmetry. However, after decades of investigation, it appears that hepatitis A virus is a picornavirus, most likely an enterovirus, although more data are needed to settle the issue. Although the nucleic acid type for hepatitis A virus has not been conclusively determined, the evidence strongly suggests that it is RNA. Hepatitis A virus also resembles enteroviruses in its 27-nm size, spherical shape, density in cesium chloride ($1.34\,g/cm^3$), and stability to ether treatment. The virus also contains four polypeptides with molecular weights similar to those of the enteroviruses. In addition, its stability to acid pH is like that of a typical enterovirus, clearly distinguishable from rhinoviruses. Hepatitis A virus will probably be recognized as enterovirus-72.

Table III. RNA-containing viruses with cubic capsid symmetry

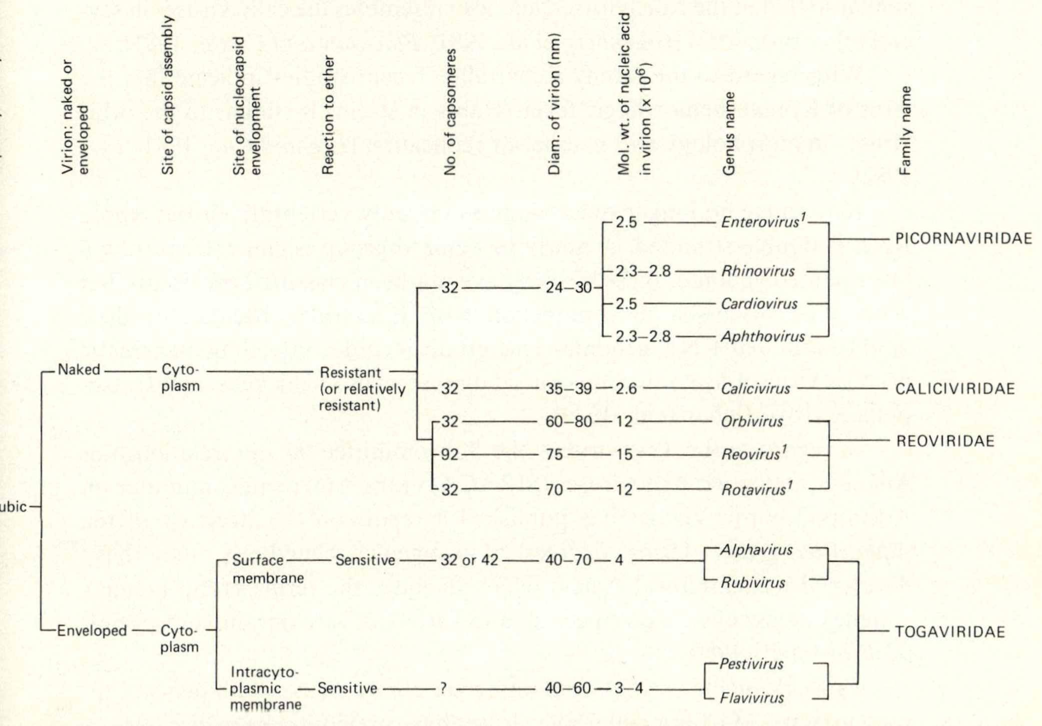

	Virion: naked or enveloped	Site of capsid assembly	Site of nucleocapsid envelopment	Reaction to ether	No. of capsomeres	Diam. of virion (nm)	Mol. wt. of nucleic acid in virion (x 10^6)	Genus name	Family name
ubic	Naked	Cyto-plasm		Resistant (or relatively resistant)	32	24–30	2.5	*Enterovirus*[1]	PICORNAVIRIDAE
							2.3–2.8	*Rhinovirus*	
							2.5	*Cardiovirus*	
							2.3–2.8	*Aphthovirus*	
					32	35–39	2.6	*Calicivirus*	CALICIVIRIDAE
					32	60–80	12	*Orbivirus*	REOVIRIDAE
					92	75	15	*Reovirus*[1]	
					32	70	12	*Rotavirus*[1]	
	Enveloped	Cyto-plasm	Surface membrane	Sensitive	32 or 42	40–70	4	*Alphavirus*	TOGAVIRIDAE
								Rubivirus	
			Intracyto-plasmic membrane	Sensitive	?	40–60	3–4	*Pestivirus*	
								Flavivirus	

Most of the RNA-containing viruses are sensitive to pH 3 treatment; exceptions are the enteroviruses, reoviruses, and otaviruses.

A Study Group report now has been published on the family Caliciviridae, recently separated from the Picornaviridae [*Schaffer* et al., 1980]. For this family, the genome is a single molecule of infectious, single-stranded, positive RNA. There is no lipid and no envelope, and there is a single major structural polypeptide rather than several as seen with the picornaviruses. Only one genus, *Calicivirus,* has been proposed; it includes a number of serotypes infecting lower animals. Calicivirus-like particles, about 37 nm in diameter, have been observed in human feces in association with gastroenteric disease. In addition, the Norwalk virus – a widespread human agent causing acute epidemic gastroenteritis – has a virion protein structure similar to that of the caliciviruses, and also resembles the caliciviruses in several other properties [*Greenberg* et al., 1981; *Blacklow and Cukor,* 1981].

With regard to the family Reoviridae, recent studies indicate that the virus of Korean hemorrhagic fever (Hantaan strain) is similar to the orbiviruses in morphology and manner of replication [*Lee and Cho,* 1981; *Lee,* 1982].

Reoviridae no longer stand alone as the only vertebrate viruses whose RNA is double-stranded. A newly recognized group is characterized by a bisegmented genome; these viruses have not been classified previously but were often discussed in conjunction with Reoviridae because of their double-stranded RNA genome. This group includes infectious pancreatic necrosis virus of fish, infectious bursal disease virus of chickens, and Drosophila X virus [*Dobos* et al., 1979].

In regard to the Togaviridae, the Subcommittee on Interrelationships Among Catalogued Arboviruses (SIRACA) of the American Committee on Arthropod-borne Viruses has published a report on the structure of the *Alphavirus* genus. Using degrees of antigenic relatedness, they have developed a hierarchical system which includes the terms group (genus), complex (subgenus), type (species), and variety or serovariant (subspecies) [*Calisher* et al., 1980].

Table IV includes RNA-containing vertebrate viruses with helical capsid symmetry. Marburg and Ebola viruses have previously been discussed as possible members of Rhabdoviridae because of some similarities. However, recent evidence indicates that the RNA and proteins of these viruses have physicochemical properties that differ from those of the rhabdoviruses.

Subdivision of the family Bunyaviridae into genera [*Bishop* et al., 1980] has been proposed to reflect both antigenic supergroup relationships and molecular similarities and differences. The genus *Bunyavirus* includes at least 124 viruses, most of them belonging to 13 serologically cross-related groups; several ungrouped arboviruses also are included; most of these agents are mosquito-transmitted. Among the members are the virus of California encephalitis, and also the La Crosse virus which causes human encephalitis with occasional fatal outcome, as well as other viruses that cause febrile diseases in man. Other bunyaviruses cause disease in some ruminants. The genus *Phlebovirus* includes the virus of Phlebotomus (sandfly) fever of man, Rift Valley fever virus, and at least 28 other viruses. These agents are predominantly sandfly-borne, but some have been recovered from mosquitoes; a wide variety of vertebrates, especially rodents, may be infected. Rift Valley

able IV. RNA-containing viruses with helical capsid symmetry

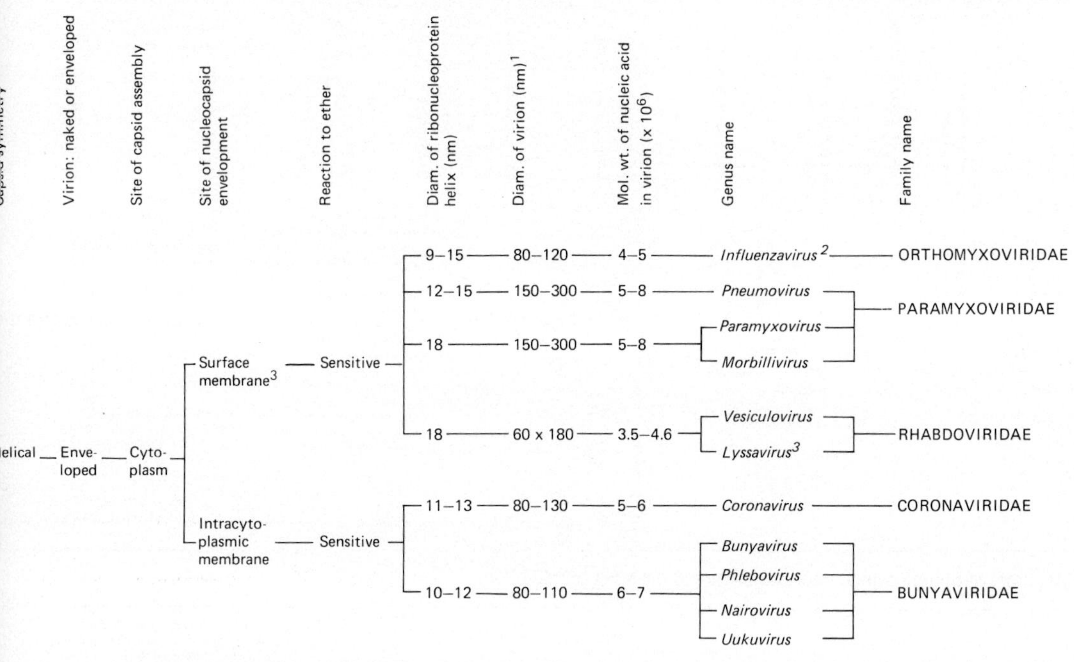

Capsid symmetry	Virion: naked or enveloped	Site of capsid assembly	Site of nucleocapsid envelopment	Reaction to ether	Diam. of ribonucleoprotein helix (nm)	Diam. of virion (nm)[1]	Mol. wt. of nucleic acid in virion (x 10^6)	Genus name	Family name
Helical	Enveloped	Cytoplasm	Surface membrane[3]	Sensitive	9–15	80–120	4–5	*Influenzavirus*[2]	ORTHOMYXOVIRIDAE
					12–15	150–300	5–8	*Pneumovirus*	PARAMYXOVIRIDAE
					18	150–300	5–8	*Paramyxovirus*	
								Morbillivirus	
					18	60 x 180	3.5–4.6	*Vesiculovirus*	RHABDOVIRIDAE
								Lyssavirus[3]	
			Intracytoplasmic membrane	Sensitive	11–13	80–130	5–6	*Coronavirus*	CORONAVIRIDAE
					10–12	80–110	6–7	*Bunyavirus*	BUNYAVIRIDAE
								Phlebovirus	
								Nairovirus	
								Uukuvirus	

Diameter, or diameter × length.
Influenza type C probably is a separate genus, but has not yet been so designated.
Rabies virus buds predominantly from intracytoplasmic membranes.

fever virus, a serious pathogen of sheep, has recently caused large epidemics among humans in Africa. The genus *Nairovirus,* named from the virus of Nairobi sheep disease, has as its prototype the more intensively studied Crimean-Congo hemorrhagic fever virus. Nairoviruses are predominantly tick-transmitted. At least 19 serotypes belong to this genus. Members of the genus *Uukuvirus* are tick-transmitted. The genus includes the type species, Uukuniemi virus, and 6 other viruses. They have been isolated in nature from rodents and from ticks, but have no known pathogenicity for man.

Table V shows the RNA-containing viruses with architecture unsymmetric or unknown. Included are the family Retroviridae and the family Arenaviridae.

Table V. RNA-containing viruses with architecture unsymmetric or unknown

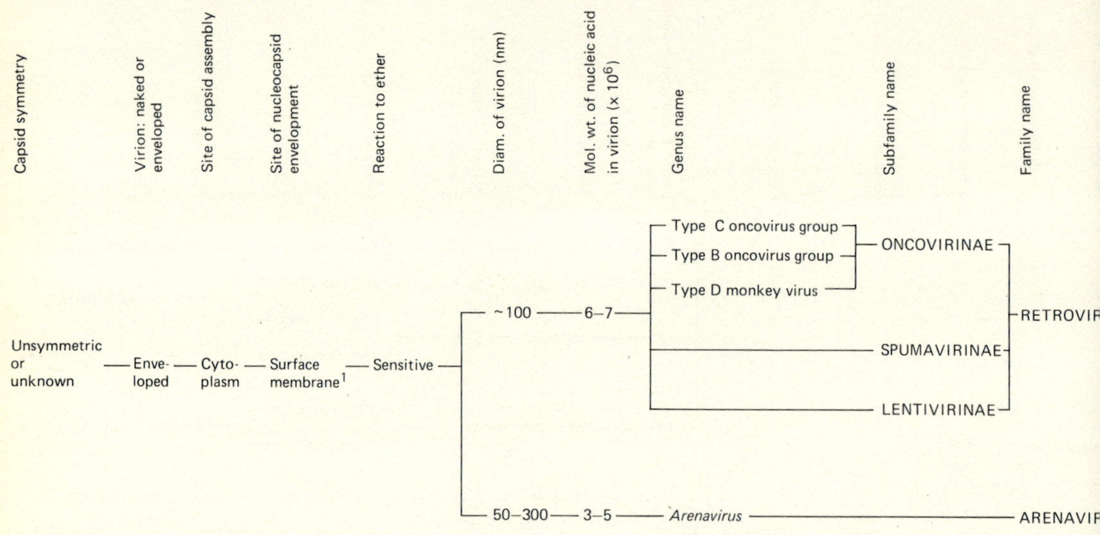

[1] Members of Spumavirinae bud into intracytoplasmic vacuoles.

Hepatitis B Viruses

These agents look very much like a virus family. Much is known about hepatitis B virus of man, but many of its properties are difficult to determine unequivocally because it has not yet been successfully propagated in cell culture. Sera from patients with hepatitis B infections reveal three distinct morphological entities in varying proportions. The more numerous forms (by a factor of more than a thousand) are the small pleomorphic spherical particles measuring about 22 nm in diameter (in concentrations as high as 10^{12} particles per ml). Tubular or filamentous forms of similar diameter but varying length also are observed. Hepatitis B virus (HBV) is a complex, double-shelled particle 45 nm in diameter, containing a 28-nm core surrounded by a lipoprotein coat 7 nm thick. Within the core is a circular, double-stranded DNA molecule and a DNA-dependent DNA polymerase. The DNA is made up of one strand of fixed length and a shorter strand of variable length.

A woodchuck hepatitis virus (WHV), found in *Marmota monax* in Pennsylvania, is a second member of this family. HBV and WHV have the following characteristics in common: (i) infection with either virus results in accumulation in the blood of large amounts of excess viral coat protein in the form of spherical and tubular particles 20–25 nm in diameter; (ii) particles 40–50 nm in diameter, containing a small, partially single-stranded circular DNA and a DNA polymerase, may also be found in blood and are thought to be the complete virions; and (iii) both viruses are associated with chronic hepatitis and hepatocellular carcinoma. The surface and core of WHV have specific antigenic determinants. Anti-HBc is present in the serum of humans with HBV infections; similarly, antibody to the core of WHV (anti-WHc) is found in WHV-infected woodchucks. The nucleocapsid components (cores) of the two viruses share major determinants; this reflects a group-specific antigen for this family.

A second animal virus (GSHV) with many of the unique characteristics of human HBV has been found in ground squirrels *(Spermophilus beecheyi)* in northern California. Common features include virus morphology, viral DNA size and structure, a virion DNA polymerase that repairs a single-stranded region in the viral DNA, cross-reacting viral antigens, and persistent infection with viral antigen continuously in the blood. GSHV is closely related to WHV.

Duck hepatitis B virus (DHBV) has been detected in Pekin ducks *(Anas domesticus)* from flocks in China and in the USA. On the basis of virus morphology, virion DNA polymerase activity, size and structure of the virus genome, and association between virus in the serum and large amounts of viral DNA in the liver, this agent is also included in the hepatitis B family of viruses.

Virus Species

Since there has been so much progress in establishing the higher taxa, the question of virus species is now approaching center stage [*Matthews,* 1979]. Recent proposals of the Study Groups on Adenoviridae and Herpesviridae serve to illustrate some of the approaches and problems.

The Adenoviridae Study Group has proposed to define its species as follows:

'A species (formerly type) is defined on the basis of its immunological distinctiveness, as determined by quantitative neutralization of animal antisera. A species has either no cross-reaction with others or shows a homologous-to-heterologous titer ratio of greater than 16 in both directions. If neutralization shows a certain degree of cross-reaction between two viruses in either or both directions (homologous-to-heterologous titer ratio of 8 to 16), distinctiveness of species is assumed if: (a) the hemagglutinins are unrelated, as shown by lack of cross-reaction in hemagglutination-inhibition; or (b) substantial biophysical/biochemical differences of the DNAs exist.'

In the recommendations, human adenovirus species would be designated by the letter h followed by the present Arabic number series (for example, mastadenovirus h 1). Adenoviruses of domestic animals would be designated by a three-letter code based on the genus of the respective host animal (for example, mastadenovirus bos 1). Species of aviadenoviruses would be named in the same way. Thus, for the adenoviruses of man, the species would be h 1 through h 34; for those of cattle, bos 1 – bos 9; for those of pig, sus 1 – sus 4; for those of fowl, gal 1 – gal 9; of turkeys, mel 1 – mel 2, and so on.

For Herpesviridae, the Study Group has recommended [*Roizman* et al., 1981] that, to provide a working nomenclature, the following provisional system be used as a formal interim step toward a universal taxonomy:

'(1) Each herpesvirus should be named after the taxonomic unit – the family or subfamily – to which its primary natural host belongs. The *subfamily* name should be applied to viruses isolated from the host family Bovidae and from primates; in all other instances the *family* name should be used. When family names are used, the name should end in 'id' (e.g., equid, caviid, etc.). Names based on the subfamily to which the host belongs should end in 'ine' (e.g., bovine, cebine, etc.). The only exceptions are viruses isolated from man; because humans are the sole living members of the family Hominidae, the term human rather than hominid shall be used...'

(2) The herpesviruses within each host family or subfamily would be given Arabic numbers... New herpesviruses would receive the next available number. An alternative designation for individual herpesviruses, to show clearly their subfamily affiliation, would include the subfamily prefix as a part of the species name. For example, the varicella-zoster virus would have the species designation 'human herpesvirus 3' or 'human alphaherpesvirus 3'; herpes simplex virus type 1 would be 'human herpesvirus 1' or 'human alphaherpesvirus 1'; human cytomegalovirus would be human herpesvirus 5 or human betaherpesvirus 5, while cytomegalovirus of mice would be murid herpesvirus 1 or murid betaherpesvirus 1, etc.

Nomenclature of Influenza Virus Subtypes

The ICTV has not dealt with matters of strain characterization and nomenclature. However, identification and clear description of influenza

Table VI. Some reference strains for the newly designated subtypes of hemagglutinin and neuraminidase antigens of influenza A viruses isolated from man, swine and horses

H and N subtypes	Reference strains
H1N1	A/PR/8/34 (H1N1)
	A/FM1/47 (H1N1)
	A/England/1/51 (H1N1)
	A/New Jersey/8/76 (H1N1)
	A/USSR/90/77 (H1N1)
H2N2	A/Singapore/1/57 (H2N2)
	A/Tokyo/3/67 (H2N2)
H3N2	A/Hong Kong/1/68 (H3N2)
	A/Port Chalmers/1/73 (H3N2)
	A/Texas/1/77 (H3N2)
H1N1	A/swine/Iowa/15/30 (H1N1)
	A/swine/Wisconsin/67 (H1N1)
H3N2	A/swine/Taiwan/1/70 (H3N2)
H7N7	A/equine/Prague/1/56 (H7N7)
H3N8	A/equine/Miami/1/63 (H3N8)

virus subtypes, particularly within type A, have long been recognized as essential for dealing with the antigenic variability of these viruses and the periodic emergence of new strains against which population immunity is low or absent. A system of nomenclature recommended by the *World Health Organization* [1980] includes a type and strain designation and a description of the antigenic specificity of the surface antigens, the hemagglutinin (H) and the neuraminidase (N). For all three types, the strain designation includes information on the antigenic type of the virus (based on the antigenic specificity of the nucleoprotein) as A, B, C; the host of origin (for strains isolated from non-human species), geographical origin, strain number, and year of isolation. For viruses of type A, this strain designation is followed by a second part, in parentheses, indicating the antigenic subtype of the hemagglutinin and of the neuraminidase antigens. For influenza A viruses from all species, the H antigens are now grouped into 12 subtypes, H1–H12; the N antigens are divided into 9 subtypes, N1–N9. Although it is recognized that antigenic variation occurs among influenza B strains, division into subtypes is not recommended.

The WHO consultative group recommended that this revised system should be used universally [*WHO*, 1980]. Table VI gives examples of reference strains from man, swine, and horses, using the new nomenclature.

Nomenclature of Poliovirus Strains

Another group of viruses for which strain differentiation is of direct concern in medical virology are the polioviruses. In 1951 the Committee on Typing of the National Foundation for Infantile Paralysis reported that polioviruses should be assigned to one of three serological groups. In the intervening years, intratypic antigenic differences were established by means of serodifferentiation tests but, despite further refinements, the scope of possible studies was limited because of difficulties in obtaining suitable strain-specific sera.

Recent work has brought new tools, highly strain-specific absorbed sera and oligonucleotide mapping procedures, to the study of poliovirus strains. A World Health Organization group of virologists has called attention to important implications of these new methods. The methods were used to show a common strain origin for the type 1 poliovirus responsible for epidemics among non-vaccinated persons in the Netherlands and Canada in 1978, and in the USA in 1979. Also, all recent type 3 poliovirus isolates in the United Kingdom were found to be related to the type 3 live poliovaccine strain. Because this increased ability to characterize new isolates serologically has epidemiological significance, a more precise system of nomenclature for poliovirus strains is needed.

Poliovirus isolates should be identified by type, country (or city), strain number, and year of isolation. Thus P1/England/119/65 designates a type 1 poliovirus strain, number 119, isolated in England in 1965. Whenever reference is made to a poliovirus strain not previously described in a publication, it should be identified by this system of nomenclature [*World Health Organization*, 1981].

Comment

The taxonomy of the viruses of vertebrates has been advanced remarkably in recent years. A particularly impressive aspect of this achievement is the wide acceptance and almost universal use of the family and genus designations developed by the ICTV since its establishment in 1966. An important contributing factor has been the active involvement of working virologists in each step of the process. The annual advancements may be gleaned from reviewing the previous chapters on viral taxonomy in *Progress in Medical Virology.* More detailed presentations that also include viruses of invertebrates, plants, and bacteria may be found in the ICTV Reports edited by *Wildy* [1971], *Fenner* [1976], and *Matthews* [1979].

References

Bishop, D.H.L.; Calisher, C.H.; Casals, J.; Chumakov, M.P.; Gaidamovich, S.Ya.; Hannoun, C.; Lvov, D.K.; Marshall, I.D.; Oker-Blom, N.; Pettersson, R.F.; Porterfield, J.S.; Russell, P.K.; Shope, R.E.; Westaway, E.G.: Bunyaviridae. Intervirology *14:* 125–143 (1980).

Blacklow, N.R.; Cukor, G.: Viral gastroenteritis. New Engl. J. Med. *304:* 397–406 (1981).

Calisher, C.H.; Shope, R.E.; Brandt, W.; Casals, J.; Karabatsos, N.; Murphy, F.A.; Tesh, R.B.; Wiebe, M.E.: Proposed antigenic classification of registered arboviruses. I. Togaviridae, *Alphavirus.* Intervirology *14:* 229–232 (1980).

Dobos, P.; Hill, B.J.; Hallett, R.; Kells, D.T.C.; Becht, H.; Teninges, D.: Biophysical and biochemical characterization of five animal viruses with bisegmented double-stranded RNA genomes. J. Virol. *32:* 593–605 (1979).

Fenner, F.: Classification and nomenclature of viruses. Second report of the International Committee on Taxonomy of Viruses. Intervirology *7:* 1–115 (1976).

Greenberg, H.B.; Valdesuso, J.E.; Kalica, A.R.; Wyatt, R.G.; McAuliffe, V.J.; Kapikian, A.Z.; Chanock, R.M.: Proteins of Norwalk virus. J. Virol. *37:* 994–999 (1981).

Lee, H.W.: Korean hemorrhagic fever. Prog. med. Virol., vol. 28, pp. 96–113 (Karger, Basel 1982).

Lee, H.W.; Cho, H.J.: Electron microscopic appearance of Hantaan virus, the causative agent of Korean haemorrhagic fever. Lancet *i:* 1070–1072 (1981).

Matthews, R.E.F.: Classification and nomenclature of viruses: Third Report of the International Committee on Taxonomy of Viruses. Intervirology *12:* 129–296 (1979).

Roizman, B.; Carmichael, L.E.; Deinhardt, F.; de-The, G.; Nahmias, A.J.; Plowright, W.; Rapp, F.; Sheldrick, P.; Takahashi, M.; Wolf, K.: Herpesviridae: definition, provisional nomenclature, and taxonomy. Intervirology *16:* 201–217 (1981).

Schaffer, F.L.; Bachrach, H.L.; Brown, F.; Gillespie, J.H.; Burroughs, J.N.; Madin, S.H.; Madeley, C.R.; Povey, R.C.; Scott, F.; Smith, A.W.; Studdert, M.J.: Caliciviridae. Intervirology *14:* 1–6 (1980).

Wildy, P.: Classification and nomenclature of viruses: First Report of the International Committee on Nomenclature of Viruses. Monogr. Virol., vol. 5 (Karger, Basel 1971).

World Health Organization Consultation: A revision of the system of nomenclature for influenza viruses: a WHO Memorandum. Bull. Wld Hlth Org. *58:* 585–591 (1980).

World Health Organization: Nomenclature for strains of poliovirus. Wkly epidem. Rec. *56:* 231 (1981).

Dr. Joseph L. Melnick, Department of Virology and Epidemiology,
Baylor College of Medicine, Houston, TX 77030 (USA)

Subject Index, Vol. 28

Cumulative Author Index

Vol. 25 (1979), Vol. 26 (1980), Vol. 27 (1981), and Vol. 28 (1982)
Cumulative Author Index for vol. 20–25 may be found in vol. 25, pp. 171–174, 1979
Cumulative Author Index for vol. 1–19 may be found in vol. 19, pp. 373–381, 1975

Cumulative Title Index

Vol. 25 (1979), Vol. 26 (1980), Vol. 27 (1981), and Vol. 28 (1982)
Cumulative Title Index for vol. 20–25 may be found in vol. 25, pp. 175–179, 1979
Cumulative Title Index for vol. 1–19 may be found in vol. 19, pp. 363–372, 1975